A Holistic Approach to Lessons Learned

A Holistic Approach to Lessons Learned

How Organizations Can Benefit from Their Own Knowledge

Moria Levy

CRC Press
Taylor & Francis Group
Boca Raton London New York

CRC Press is an imprint of the
Taylor & Francis Group, an **informa** business

AN AUERBACH BOOK

CRC Press
Taylor & Francis Group
6000 Broken Sound Parkway NW, Suite 300
Boca Raton, FL 33487-2742

© 2018 by Taylor & Francis Group, LLC
CRC Press is an imprint of Taylor & Francis Group, an Informa business

No claim to original U.S. Government works

Printed on acid-free paper

International Standard Book Number-13: 978-1-138-56476-3 (Hardback)

This book contains information obtained from authentic and highly regarded sources. Reasonable efforts have been made to publish reliable data and information, but the author and publisher cannot assume responsibility for the validity of all materials or the consequences of their use. The authors and publishers have attempted to trace the copyright holders of all material reproduced in this publication and apologize to copyright holders if permission to publish in this form has not been obtained. If any copyright material has not been acknowledged please write and let us know so we may rectify in any future reprint.

Except as permitted under U.S. Copyright Law, no part of this book may be reprinted, reproduced, transmitted, or utilized in any form by any electronic, mechanical, or other means, now known or hereafter invented, including photocopying, microfilming, and recording, or in any information storage or retrieval system, without written permission from the publishers.

For permission to photocopy or use material electronically from this work, please access www.copyright.com (http://www.copyright.com/) or contact the Copyright Clearance Center, Inc. (CCC), 222 Rosewood Drive, Danvers, MA 01923, 978-750-8400. CCC is a not-for-profit organization that provides licenses and registration for a variety of users. For organizations that have been granted a photocopy license by the CCC, a separate system of payment has been arranged.

Trademark Notice: Product or corporate names may be trademarks or registered trademarks, and are used only for identification and explanation without intent to infringe.

Visit the Taylor & Francis Web site at
http://www.taylorandfrancis.com

and the CRC Press Web site at
http://www.crcpress.com

To my beloved husband who always inspires
me to do more and better

Contents

FOREWORD	xi
PREFACE	xv
ACKNOWLEDGMENTS	xxi
AUTHOR	xxiii

PART I THE NEED

CHAPTER 1	A WORLD FULL OF CHALLENGES	3
	The Role of Knowledge	3
	Competition	5
	External Information	6
	Internal Information and Knowledge	7
	References	12

PART II CREATING THE KNOWLEDGE

CHAPTER 2	LEARNING THE LESSON FOR THE FIRST TIME	15
	When Should We Debrief?	16
	Who Should Debrief?	21
	How to Debrief?	23
	References	23
CHAPTER 3	DEEPENING: DEBRIEFING TECHNIQUES AND GUIDELINES	25
	Techniques of Debriefing	26
	After Action Review	26
	Enlarging the AAR Method	29

		Multi-Case Learning: Learning from Several	
		Processes or Events	33
		References	37
Chapter 4		**Learning from Quality-Based Processes**	39
		Learning from the Plan–Do–Check–Act Model (PDCA)	40
		Learning from the Define, Measure, Analyze, Improve, and Control Method (DMAIC)	41
		Learning from Gemba Walks	42
		Learning from Quality Audits	43
		Quality-Based Processes	45
		Conclusion	45
		References	46
Chapter 5		**Experience**	47
		The Process of Experiential Learning: Kolb's Four Forms of Learning	48
		The Process of Experiential Learning: Kahneman and the Two Systems Theory	50
		The Process of Experiential Learning: Cell's Four Levels of Experiential Learning	51
		Integrating What We Learned	53
		References	56

Part III Managing Created Knowledge

Chapter 6	**Actions**	59
	References	63
Chapter 7	**Knowledge and a Knowledgebase to Handle It**	65
	Saving the Lessons in a Knowledgebase	69
	The Structure of the Knowledgebase	73
	The Lessons	77
	Reference	80
Chapter 8	**Embedding the Lessons in the Organizational Environment**	81
	Templates and Forms	82
	Training	84
	Making It Hard to Make the Mistake	86
	Checklists	88
	Presenting Lessons in Search Processes	89
	Online Help	90
	Processes	91
	Work Procedures and Guidelines	92
	Summary	93
	References	94

Part IV Returning to the Previous Day

Chapter 9 Requesting the Knowledge before Action — 97
    BAR: Before Action Review — 101
    Reference — 104

Part V Implementing the Life-Cycle Model of Lessons and Good Practices Management

Chapter 10 Jumping into the Water — 107
    The Decision to Start — 108
    Scope — 110
    Choosing with Whom to Start — 111
    Determining the Order in Which to Implement Change — 112
    Summary — 113
    Reference — 114

Chapter 11 The Cultural Change — 115
    Initiating the Move — 118
    Helping People through the Journey of Change — 119
    Summary — 122
    References — 122

Chapter 12 Roles in the Organization — 123
    The Sponsor — 124
    The Lessons Manager — 125
    The Knowledgebase Manager — 125
    Subject Matter Experts — 127
    Employees — 127
    Reference — 127

Chapter 13 Summary — 129
    Surprises — 129
    Debriefing Techniques — 130
    Additional Sources — 130
    Outcomes — 131
    Closing the Loop — 132
    Glossary: Review of Terms — 134
        AAR: After Action Review — 134
        BAR: Before Action Review — 134
        Best Practices (Good Practices) — 135
        Debriefing — 135
        Knowledgebase — 136
        Knowledge Management — 136
        Learning — 137
        Learning from Experience — 137
        Learning from Quality-Based Processes — 138

Learning Organization 138
　　　Lessons/Lessons Learned 138
　　　　Team Learning 139
　　References 139
INDEX 141

Foreword

"It may metaphorically be said that natural selection is daily and hourly scrutinizing, throughout the world, the slightest variations; rejecting those that are bad, preserving and adding up all that are good; silently and insensibly working, whenever and wherever opportunity offers, at the improvement of each organic being..." (Charles Darwin)

Darwin called the principle, by which each variation is preserved if useful, *natural selection*. Nature, then, was the first to implement the process of learning and debriefing.

Of course, the process of learning and debriefing is nothing new to human civilization. Man and mankind have always been constantly involved in learning and improving; using what worked and searching for alternatives for failure. Thus, whether by a methodological process or by intuitive thinking, humanity has evolved through learning and debriefing. Yet it is humanity's rapid evolution, specifically the rate of changes, technological developments, the accumulation of data and knowledge, and the quick flow of information that characterize the twenty-first century. These have placed society in a race for survival, in which the victors are those who can learn and adapt better and faster than others.

The process of learning and debriefing, therefore, is essential for any organization that wishes to survive, and certainly for those that wish to succeed and prosper. Just like Darwin's principle of natural selection—the fittest (and the adaptable) survive and progress.

Why, in fact, do organizations aspire (and are obligated) to learn and improve? One can come up with a number of important reasons: First, to avoid past mistakes. But more than that—an efficiently learning organization can reach greater success the first time. Such a process provides the organization and its employees with confidence in their performance, improves their ability to recover from failure, and enhances their organizational excellence.

However, it turns out that efficient implementation of a learning process is not a simple matter. First, since this process involves, by definition, dealing with mistakes, it is natural that people do not find it easy to develop an open and revealing conversation regarding their mistakes. Thus, implementing a true learning process touches on the foundations of organizational culture; the same vague definition for "how we do things around here" that enables and encourages people to discuss their shortcomings, through seeing the clear benefit to the entire organization.

Even in a culture that embraces a thorough and revealing debriefing process, the challenge of incorporating these lessons into the organization and turning them into practice (i.e., changing and improving the way in which the organization performs) still remains. This calls for an apt methodology. Albert Einstein once said that "we can't solve problems by using the same kind of thinking we used when we created them." Implementing lessons in organizations requires a methodology that at the end of the day leads to a change in the way we act.

One organization in which a meticulous and successful process has developed and implemented is the Israeli Air Force (IAF), where I have served for 37 years, four of them as its commander. The roots of the IAF's debriefing culture stem from its critical need in the world of aviation, which promotes excellence and mistakes can carry a heavy price, and in which every detail is vital. This basic need for debriefing in the world of aviation led to the growth and development of a culture that enables, appreciates, and even encourages the openness to bring up mistakes and shortcomings on all levels, to analyze them using a scalpel, and to reach conclusions and improve. The IAF also thoroughly implements a systematic methodology for debriefing and learning lessons.

These two components, organizational culture and the methodology, are the cornerstones of any adaptable organization. Regarding the IAF, one can clearly argue that its high quality is derived explicitly from its ability to learn and to reach conclusions in all areas in which it engages, based on its systematic methodology and its unique organizational culture.

Once the need is clear, how do we implement a healthy learning and debriefing process in a manner that will allow us to successfully compete in a rapidly changing world? First, we must address the aspects related to organizational culture in order to allow an open discourse regarding shortcomings and mistakes. It is important to note that organizational culture differs between organizations, and an attempt to replicate it fully might fail. It is therefore important to carefully examine and select the aspects desired for the process of learning and debriefing, and to implement them in a manner that fits our organization. It is worth stressing that the implementation of changes within an organizational culture, and certainly the encouragement of open discourse, begins with the organization's managers, who set the obligated standards, tolerance, and integrity.

Regarding methodology—without an orderly system, lessons will neither be collected nor implemented throughout the organization. Therefore, the system in which conclusions are drawn and implemented is to be defined and made a part of the natural work mechanisms of the organization.

There is no definite recipe, nor can one be made, for the right organizational culture or a winning methodology. As previously mentioned, even if one encounters an organization that successfully generates a culture and methodology for lesson learning, it is obviously not possible to replicate it as is. Nonetheless, there is definitely a call for an in-depth discussion over different perceptions of culture and methodology, to treat them as an all-encompassing outline, and to perform the appropriate adjustments for our organization. This way, it is possible to gradually reinforce good patterns, to add processes, and to make learning a part of the organizational DNA.

In her book, *Holistic Approach to Lessons Learned: How Organizations Can Benefit from Their Own Knowledge*, Dr. Moria Levy suggests a methodical and organized way of elevating the lesson learning process to organizations that wish to improve their ability to successfully compete and succeed, and offers the tools and mindset to improve it.

It is possible and beneficial to learn from the experience of others how to learn and implement lessons.

Ido Nehutshtan
Major General (retired)

Preface

The chief executive officer (CEO) of X2X, Inc. moved uneasily in his chair. He had just been informed that the company's newly developed phone had not passed its acceptance testing. Clients who had tried out the new model had complained that the device emitted heat. The research and development (R&D) vice president (VP) updated him that they had found a design flaw. *The hell with it,* thought the CEO. *We know what is wrong. We previously had encountered that same problem in our model developed 2 years ago. Then, too, we detected a design flaw; lessons had been learned. How are we back at square one 2 years later? How was a lesson, previously learned, totally ignored, forcing us to learn it once again from scratch?*

Confucius, a Chinese philosopher and thinker, stated nearly 2000 years ago: "If you make a mistake and do not correct it, this is called a mistake." (Confucius) To paraphrase, it can be said: "While it is not a mistake to err, it is certainly an error to repeat one."

Why, then, do we repeat past mistakes, both as individuals and as organizations?

Let us start at the very end. Not all mistakes can be prevented. Occasionally, a malfunction seemingly reoccurs, yet although it features a similar element, the situation is altogether different. An organization offering services can lose a tender twice, each time for a different reason. It is impossible to totally prevent the recurrence of errors and mistakes. That said, in many cases, mistakes can and should be prevented from repeating themselves.

Many organizations deal with organized lesson-learning processes to learn from past mistakes and in turn to improve future performance. Many of these lesson-learning methods regarding events or activities can be performed in small or large groups; these methods involve, in addition to failure analysis, success analysis for purposes of reproduction.

Nevertheless, organizations err and repeat mistakes.

This book offers an expanded approach that enables a substantial reduction of recurring mistakes and lessons repeatedly learned. This approach, developed by the author, is based on knowledge management methods and the author's experience in lesson learning and knowledge management in organizations for more than 20 years. This approach has been successfully implemented in various types of Israeli organizations: business and public; small, midsize, and large; and veteran and new organizations. Field experience shows that implementing this method indeed reduces recurring mistakes and redundantly repeated lessons and therefore improves organization performances.

This approach is based on implementing a life-cycle model of lessons and good practices management. Efficient and effective lesson production is merely one piece of the entire lesson management. Lesson management includes four central stages: (1) creating new knowledge; (2) processing and distilling said knowledge while separating it from tasks and related changes; (3) incorporating knowledge into the current organizational environment; and (4) finally, reusing the knowledge before the next event, process, or action.

The book's structure reflects this life-cycle model of lessons and good practices management:

Chapter 1 discusses the challenges we face that have made the need for lesson learning and management even greater than before.

Chapters 2 through 5 discuss ways to create knowledge. Chapter 2 discusses managing lesson production processes, and Chapter 3 delves into methods that can be implemented in any organization to effectively and efficiently produce lessons. Chapters 4 and 5 expand the sources of new knowledge that are recommended as part of the overall lesson management framework; Chapter 4 reviews complementary learning tools owned by many knowledge-generating organizations.

Often, although organizations performed processes such as quality surveys, customer feedback management, and such, these are managed apart from the lessons; this book offers a holistic approach that merges new knowledge acquired via these processes with knowledge gained through lesson learning. Chapter 5 contributes yet another source of knowledge: Experience. Experience is created in the organization without any proactive activities, yet few organizations manage it regularly and fewer manage it in conjunction with the lessons.

This concludes the review of knowledge sources.

Chapters 6 and 7 are related to processing knowledge and separating it from tasks and changes. Chapter 6 discusses immediate long-term actions that must be performed in response to the created knowledge, short-term tasks, and medium-term change management. In many cases, these are all mixed together and their lack of differentiation harms everyone as the correct handling of lessons, tasks, and changes differ greatly. Chapter 7 expands on managing the created knowledge. This chapter defines processes supporting the refinement of this knowledge to be sufficiently beneficial and suggests a method for managing accumulated lessons in a unique knowledgebase. The approach suggested in this chapter is part of the innovations of the general approach of this book because usually only lesson management documents are managed, yet the actual lessons do not undergo any processing.

The life-cycle then advances and reaches its third stage: incorporating the knowledge into the organizational environment. Chapter 8 discusses methods and tools for this merging process. As in many other areas of life, there is no uniform recipe for incorporating knowledge into the organizational environment, which is applicable for all organizations and all lessons. This chapter, therefore, offers various solutions that either can be implemented exactly as cited or utilized as a basis for other ideas relevant to the specific organizational context.

Chapter 9 discusses the final stage, or more accurately, the first stage in the lesson management life-cycle. This chapter discusses methods that enable workers to acquire new knowledge and choose a course of action that implements this knowledge before an activity, event, or process.

This chapter highlights the unique approach suggested in this book: Managing an entire lesson life-cycle that includes the creation of the knowledge, designated management of the knowledge items, merging them into the organizational environment, and taking measures to encourage others to utilize said knowledge. This approach substantially increases this knowledge's chances of reaching a worker requiring it in a focused, context-related manner. It therefore increases the chances that a mistake will be repeated only once.

This book also includes three final chapters that actualize the lesson management approach in organizations.

Chapter 10 deals with the starting point. The suggested approach is comprehensive: no new approach is easy to implement, especially not one that includes many components. This chapter suggests how to open the organizational door in which managers already are encumbered and persuade them to allow such a project in their organization. The chapter suggests the initial stages of this attempt, also emphasizing which stages should be postponed.

Chapter 11 discusses organizational change management and the various ways it can be implemented. The issue is not unique to this field, yet its characterization greatly affects change management methods. Chapter 12 defines participating officials whose cooperation is critical to successfully managing lessons in the suggested life-cycle.

The last chapter, Chapter 13, returns to the macro level, the big picture. It reviews the life-cycle model of lessons and good practices management (including all its components) while highlighting the main innovation suggested by this approach, as compared with other known lesson-managing approaches currently used by organizations.

Who is this book's target audience?

We all learn lessons, whether formally or in other ways. Therefore, anyone can benefit from this book. Nevertheless, this book is mainly targeted at an organizational audience. It is suitable for the following: CEOs and executives wishing to reduce the number of recurring organizational mistakes and promote lesson-learning processes; quality, knowledge, and HR managers responsible for lesson learning in their respective organizations; managers wishing to implement the approach in their unit, project, or team; and workers wishing to improve themselves and others.

A personal note: As of this writing I have been involved in the field of knowledge management for nearly 20 years. I have dealt with many knowledge-related fields. Yet, I have always felt that this area of lessons learned has something special about it. The need is crystal clear; so is the potential benefit that can be attained by implementing the full life-cycle model of lessons and good practices management approach. There are few fields in which the return on investment is as significant. I have seen organizations implement this method. The results speak for themselves.

Reference

Confucius. *Quotes*. http://www.goodreads.com/quotes/1908-if-you-make-a-mistake-and-do-not-correct-it (accessed February 16, 2017).

Acknowledgments

Writing this book was a project. One can say it took 3 years, and this indeed is the time of the writing and editing itself. But the book and its ideas started earlier; much earlier. From the day I started thinking as an adult.

First, I owe it all to my parents, Chava and Jehuda Locker, who were both pioneers in their careers (diagnostic and improvement of learning disabilities through games and human factors engineering, respectively). They taught me that I can dream and fulfill any new idea I wish, developing any new discipline that I understand is required. That is what they did. They started their journey 60 years ago.

Fast-forwarding to the year 2000, I wish to thank Irit Milo, one of my first employees in ROM Knowledgeware, who developed together with me the first models of all that was presented in this book. The following years many additional employees at ROM (which I run up till today) contributed to the refinement of this model, each adding a brick or making the model better and smoother to implement in organizations.

And the list does not end here: my son-in-law Oren Hirschhorn helped out with the English and his thoughts in between, Moshe Ekroni from Verint opened my eyes as to the ideas in Chapter 4, and many other good people on the way were always there to ask, suggest, and try the new ideas to see if and how they work.

But, the most important: endless thanks to my family:

My four children: Or, Sapir, Tomer, and Kfir, who did not always have an available mother and yet are always patient with me and my ideas, sometimes coming up with helpful insights. And finally, my beloved husband Ran—with whom I go through this journey of life, work, and everything in between. Without him, this book would not be here today.

I owe you all.

Moria Levy
Israel, October 2017

Author

Dr. Moria Levy is the CEO and owner of the largest Israeli Knowledge Management firm—ROM Knowledgeware. She is also working as a researcher and lecturer in the information sciences department of Bar-Ilan University in Israel, and as an analyst and business consultant expert in the field of Knowledge Management.

Levy serves as the chair of the Israeli Knowledge Management forum, and as the chair of the ISO Knowledge Management committee in charge of defining a Knowledge Management ISO standard.

Dr. Levy completed her BSc in mathematics and computing sciences from Bar-Ilan University in 1984, MSc in computing sciences from Bar-Ilan University in 1990, and PhD from Bar-Ilan University in 2010. Her PhD thesis deepened research on the issue of Knowledge Management utilization in organizations.

Dr. Levy has published nine papers in reputed international journals and academic books. Her areas of interest as a knowledge management expert include knowledge retention, lessons learned, change management, knowledge captures and structuring, knowledge development, and social media serving Knowledge Management.

Levy developed the Knowledge Management curriculum for Israeli high schools, including a module on learning from lessons.

PART I
THE NEED

1

A World Full of Challenges

Welcome to the world of business in the twenty-first century. As with many things in life, this business world is misleading. It may seem like the same world and the same rules as those we followed 20 years ago, back at the end of the twentieth century, but it is quite different. New challenges have emerged, yielding the necessity to continually learn—on an almost-daily basis—to survive and excel. Sticking to outdated solutions (successful as they might have been) is a downfall to be avoided.

The Role of Knowledge

The first important issue we must recognize when analyzing these new conditions is that we live in an era of knowledge. Knowledge makes the world go round. Businesses perform better based on their knowledge and how they use it to leverage their performance. Organizations strive to know and to succeed based on what they know.

This was not how things were in the twentieth century. In the 1960s and 1970s, companies were preoccupied by the question of how to *lower prices* based on process efficiency. Companies invested in analyzing each process, understanding which subprocesses were included, and developing optimal ways to handle each of these specific subprocesses. Whoever could optimize better could lower costs and thus sell more. Every company had an organization and methods unit, which was in charge of defining the processes, determining how many employees were needed, and how much time were required for each process.

At the dawn of the 1980s, a new trend arrived. It no longer sufficed to (only) be cheap; customers were seeking higher *quality*. The standard of living had increased, and companies were competing to provide higher quality goods. In the 1980s organizations spoke in terms of quality assurance and Total Quality Management.

Quality assurance units were implemented everywhere and quality audits frequently took place. The National Institutes of Standards and Technology defined—in addition to regulations—a series of quality standards, and organizations followed.

The 1990s were different. Customers still wanted low prices, but they were willing to pay more for additional quality and functionality. The most important aspect became *time*. Companies understood that "time is money." By this I am not referring to the original meaning of the phrase (seeking efficiency) but rather to a new interpretation—that is, the first to the market, leads it. In pharmaceutical companies, this always was the case. Whoever came out with a unique drug first blocked others; therefore, time was always precious. Most organizations, however, could afford to take their time, thoroughly considering new products, debating what was right to develop, and how to develop it. This was no longer the case. During the 1990s, agile and spiral modes of development emerged. What was most important was to be "out there" first, or at least as soon as possible. Many products were announced and shortly later disappeared. It felt like a revolving carousel in an amusement park, spinning faster and faster, with no time to breathe.

Toward the end of the 1990s, commercial Internet burst into our lives. Now price, quality, and time were not enough. *Information*, and later *knowledge*, became the new commodities enabling organizations to excel.

The term "knowledge workers" was coined by Peter Drucker in *Landmarks of Tomorrow* (1959). Knowledge workers were defined as employees whose success was affected significantly by knowledge. Examples of such employees included architects, consultants, physicians, and teachers. In the following years, the notion of knowledge workers was elaborated. Drucker taught us all in *The Practice of Management* (1954) that organizations will succeed if they manage their processes effectively. In 1999, writing in *Management Challenges for the 21st Century*), Drucker introduced us to yet another new paradigm of management and changed our basic assumptions about the practices and principles of management. He stated the success of organizations—both profit and nonprofit organizations—will depend on the efficiency of management of the knowledge workers (Drucker, 1999). At that time, 25% of employees were considered knowledge workers.

As of 2017, nearly two decades later, the majority of employees working at an average organization are considered knowledge workers—whether bankers, carpenters, or sales representatives. Knowledge is a key component in the success of almost every role in the organization.

The conclusion is clear: knowledge is significant to employees' success. Thus, knowledge is significant to organizations' success.

Competition

Even with this clear conclusion, life is not as simple as one would hope. One could assume that the rules of the game have not changed and that the competition to which we are accustomed has only slightly changed. We might think that knowledge becoming the excelling factor was the only major change organizations had experienced. Well, that would just be wishful thinking. Cyber technology has changed our connectivity. "The world is a global village," states a familiar expression. In his book *The Lexus and the Olive Tree: Understanding Globalization* (2000), Thomas Friedman emphasized this connectivity's significance and introduced us to new meanings and implications of this togetherness, once unfathomable. In this new world, geographic location is far less important than it used to be. One can get an X-ray in one country, and someone on the opposite side of the globe can examine the films within minutes, and then send back the medical analysis signed by a doctor who does not even speak the same language as the patient.

Another example of this global connectivity may be demonstrated in the following scenario. Someone calls the customer service of his or her cellular phone, experiencing a problem. On the other side of the phone, a service representative answers. That representative might sound local to the extent of using a familiar accent, which he or she has practiced especially for the job. In reality, however, that representative may be answering the call from the other side of the globe.

Starkly contrasting this scenario with 20 or 30 years ago, twenty-first-century knowledge work also is off-shored. Digitally purchased goods can be provided, in many cases, from everywhere to anywhere. In the past, only large stable companies participated in off-shore business. Nowadays, every organization that can get the job done, and has a convincing website, is relevant.

On the one hand, this could be an advantage, as we can sell our goods to far more customers; we no longer are in a "small red sea, rather in a new big blue ocean" (Kim and Mauborgne, 2005). On the other hand, we are not out here alone, and this ocean (or globe) is full of competitors, all claiming to be the best in the field. One could assume that this change is a minor one, as the market and competitors seem to have emerged in the same ratio. But this is not the situation. The digital changes enable new competitors to join in. These may be small companies working from a home office two blocks away; a company from a developing country that had sold its solution for years, but had suffered from small markets before globalization; or even three teenagers working out of a basement. In these new terms, it can be any company that can offer us what we request.

What is clear is that the world of providers is crowded. Competition is almost impossible, and excelling is more important than ever. Customers have alternatives. Knowledge is the key to excelling in twenty-first-century markets.

External Information

We have established that knowledge is a major factor to achieving success, and because the market is crowded, excellence is a hard goal to achieve. Organizations have to excel in using information and knowledge to continue selling services and goods in a world that is no longer as comfortable for suppliers. In this environment, managers have to ask themselves how the organization can provide its employees the information and knowledge they need to perform their jobs most effectively. The answer to this question has two components: external information and internal information. We will start by analyzing the external information.

Without a doubt, the Internet is changing our lives. We are in the middle of an information revolution, and it is therefore not easy to analyze the change. The revolution's characteristics are not yet clear. Nevertheless, we are already feeling the impact of this tsunami. Studies teach us that the information found on the Internet (and not only there) doubles every 4 years! In this era, more is written and documented than we ever could have dreamed, and a lot of information on the web is shared and can be learned. Some old wise (and unknown) man once said that as the hunger problem that troubled the world in the dawn of the twentieth century has been replaced by obesity problems at the end of the century, so too has the lack of data and information that characterized the 1960s morphed into a real concern of data and information overflow by the end of the twentieth century.

This problem, however, does not seem as severe as we experienced it even 10–15 years ago. More concisely, it might not even be a problem. Do not confuse the issue: information and data continue to double in quantity every 4 years. What has changed is accessibility. Technology has changed the rules. First, communication is much faster. We no longer have to wait (seconds that seemed like minutes that seemed like an eternity) to access information on servers on the other side of the globe. Second, and more important, search engines have become much smarter. Search engines, developed by Google and others, combine human intelligence with advanced mathematical algorithms, yielding improved search results. For years, search engines included only advanced algorithms, trying to predict the more relevant and interesting results people will want to access before others (the professional term is *ranking*). Now, when these algorithms rely on user behavior, accessibility has improved dramatically, and external information is accessible. We may and can reach it, and we are obligated to use it wherever possible, improving our job performance.

Internal Information and Knowledge

For a few seconds, some may have toyed with the delusion that access to the Internet had washed our problems away. Obviously, this is not the case. Two main things still are missing: internal information and knowledge.

Why do we need internal information if such a vast amount of external information is available on the web? We need it because context matters. We need it because this specific information is relevant to our specific organization that has specific services and provides and performs in a specific market and under specific conditions. Furthermore, the information presents us as a unique organization. Information available on the Internet may support the working processes, yet it is in the wider circle. The *core information*, in most cases, is inside the organization.

Knowledge, however, is a different story. The external world does contain information, even if it cannot serve as a unique source, complementing the existing internal information. The web, however, contains much less knowledge. Even in cases in which it includes best practices and know-how, it lacks context, and therefore, it ultimately lacks the ability to effectively use these practices and know-how in action.

This is an organizational issue, and the organization must provide its employees the knowledge that is critical for job performance and success.

So where is this knowledge? Does it even exist? Is it tacit or implicit? What is its nature? How can it be located, and how can it be useful and accessible to the knowledge worker?

Review the knowledge you wish your organization would hand you. Think about the things that would help you perform your job better. What kind of knowledge would you wish to have? Now think about this for one more minute. What percent of this knowledge would you get if you requested it? What portion of this knowledge does your organization even have? What is documented? What is shared? What is easy to access?

Although organizations differ from one another, most organizations' answers to these questions do not. Every organization holds all sorts of knowledge. There is knowledge that is known only to a specific employee or team. In some cases, the knowledge exists as it was documented at the end of some project or after a debriefing session. In other cases, the knowledge was developed through an International Organization for Standardization or Capability Maturity Model Integration effort; auditors might have found that some task was not performing correctly, and after analyzing and discussing the problem thoroughly, those auditors (or maybe the team) found a way to improve the process. And so, new knowledge was developed. In other cases, people have the knowledge and even use it, yet it never has been articulated. This is typical when knowledge has been learned during the work process and remains tacit; the employees, in some cases, are not even aware of the reasons they perform in a particular way, or why they do not choose some other path for a specific task. Time after time, we perform tasks, and we may not even be aware of the knowledge we have regarding this task. Of course, employees also hold relevant knowledge that they achieved in a former job, or former organization, and that now serves them at their current job. This knowledge is not always shared.

Above all, there is the potential knowledge: the knowledge we could have acquired had we not missed the opportunity. Most organizations, on too many occasions, do not elaborate on what has happened so that they can analyze and learn how to improve their performance next time. Also, if they unexpectedly succeed, learning can take place, enabling organizations to over perform again in the future. Nevertheless, this is a rare phenomenon.

So, how come organizations stop and learn? Do we want to excel? Of course we want to succeed. Yet there are many reasons why we do not study our lessons, develop the organizational potential knowledge, and prevent similar problems from recurring. The most popular reason is time. To be exact: lack of time. We are always in a rush, performing the next five tasks that we tried to accomplish yesterday and still cannot find the time to start, end, or advance. We tend to prefer the most urgent tasks at every given moment, thus leaving less time and attention for important, future-oriented tasks. Developing new knowledge certainly falls into the "more important but less urgent tasks" category. The point is that less knowledge for us is less knowledge for the organization. Sadly, we have fewer tools to better our performance in this age of competition than we should and could have.

Furthermore, even when we *do* debrief and develop the knowledge, we tend to only partially use it, failing to utilize the insights that we already have gained. After action reviews, among other types of debriefing sessions, are held with naturally limited attendance. Following the session, a report or a slide presentation is sent out to all who attended and other relevant stakeholders. At this stage we might even feel satisfied. Knowledge was developed efficiently and shared. The next stages leave many less optimistic. Say three groups get the e-mail containing the lessons learned. The first group consists of the people who attended the debriefing session. Only a minority of these will bother opening the e-mail because it is just one less e-mail to read. In some rare cases, they will file the report and presentation somewhere, probably in the context of the project, event, or process it had to do with. The second group includes those employees who were not invited to join the session. In some cases, they will briefly skim through the documents, many of them not fully comprehending the bullet points or contents. In many cases, the e-mail will be transferred to the "trash" file without even being read as people are overloaded with e-mails and thus attempt to avoid them. Some may procrastinate, keeping the e-mail in some random folder. The result, however, is identical in all cases: the content will remain unread. This leaves us with the third group. This last group includes those who were invited, but who did not attend. These are the only people who might read the materials sent!

Yet no matter how we look at it, the majority of employees at the organization will not be exposed to the contents—that is, to the new lessons learned.

Those exposed to the material, whether by attending the meeting or by reading the material, are most likely to implement the lessons learned. But will they actually *use* the lessons learned? Not necessarily. The obstacles that prevent willing employees from using the lessons can be divided into two main groups. The first obstacle is rather simple: people tend to forget. Unless we personally were involved in the project, event, or process, many of us tend to hear news but not assimilate. Most people have to take a proactive role (physically or emotionally) to assimilate. If we were not part of it, or not truly affected by it, we may hear, we may even listen, yet still not really store the information somewhere in our brain so we can reuse the new knowledge in the future. Of course, there are also cases in which someone was involved and took part in developing the lessons and still forgot. Think for a moment: how many times have you tried to remember the solution for some problem regarding software or hardware? You are convinced that you have been in the exact same situation before; you remember the setting as well as your frustration, but you cannot recall the solution.

The second obstacle has to do with the context. At times, we know and remember, yet still do not use the knowledge. This happens when in our minds the knowledge is related to a different context. Edward Cell, writing in *Learning to Learn from Experience* (1984), explains how our brain stores, processes, and retrieves huge amounts of data, information, and knowledge. Our brain uses templates that represent the stored knowledge: every situation is compared to an existing template deemed most relevant and then analyzed. The same happens with new lessons learned in the organization. We tag them according to the context in which they were developed. Later, when these lessons are relevant, in a different situation and context, we will find it difficult to retrieve the lessons and use them. The following example may emphasize "the context problem": 10 years ago, I gave a lecture at a conference on debriefing. I was scheduled to speak in the middle of the day, after three debriefs were already presented. "Did anyone notice," I asked as I started, "that lesson number seven of the first debrief, dealing with a successful marketing process, was identical to lesson number four, two sessions later, dealing with an engineering project failure?" Not one person raised their hand. People did not make connections between the lessons because they were learned in different contexts. The audience was silent. You could say I was lucky to have had these two debriefs presented on the same day. Lucky or not, I have found many similar cases in every organization in which this issue was raised. The reasons for this included change of context, but not just that.

If you want to try it yourself run a simple audit. Select 30–40 debriefing reports, and copy the list of recommendations and lessons you find in each debrief into a new document. Go through the list and count how many recurring lessons the organization has learned. Try to analyze whether these occurred with different people, with different units, or in different contexts. You probably will find an assortment of recurring items.

As described, the reasons for this may vary. In some cases, people just do not know that someone before them has learned the same lesson; in other cases, they do not remember; and some people just do not make the connection between the situations and therefore fail to perform even though accessible, relevant organizational knowledge exists.

Knowledge is the key to success. Achieving success in the twenty-first century has become increasingly complicated. We have to manage our knowledge—specifically our lessons—better than in the past to continue to excel. It is essential.

References

Cell, E. *Learning to Learn from Experience.* Albany, NY: State University of New York, 1984.

Drucker, P.F. *The Practice of Management.* New York: Harper & Row, 1954.

Drucker, P.F. *Landmarks of Tomorrow.* New York: Harper & Row, 1959.

Drucker, P.F. *Management Challenges for the 21st Century.* New York: Harper Business, 1999.

Friedman, T.F. *The Lexus and the Olive Tree: Understanding Globalization.* New York: Farrar Straus and Giroux, 2000.

Kim, W.C. and Mauborgne, R. *Blue Ocean Strategy: How to Create Uncontested Market Space and Make Competition Irrelevant.* Cambridge, MA: Harvard Business School Publishing Corporation, 2005.

PART II
CREATING THE KNOWLEDGE

2
Learning the Lesson for the First Time

An interesting, yet not surprising, fact is that you cannot rehearse lessons if you never learned them in the first place. Organizations initially have to decide to learn the lessons. They have to decide that they *want* to learn. Such a decision can be initiated by executive management, by middle-level management, or by the knowledge workers. Lesson learning can occur at one or all three of these levels. Yet the learning must commence to enable business improvement.

To many, lesson learning is not necessarily a natural process. This is because it is seemingly unnecessary. The company can progress without it undisturbed and unharmed (or so it seems). If a child wants to leave the house and go play, and the front door of the house is closed, the child cannot fulfill his or her wishes without opening the door. He or she will learn to open the door as part of the journey. Yet, after walking out the door, what reason does he or she have to stop and shut the door before continuing down the street and off to the slides and swings? No reason whatsoever, according to a child's logic. The way is clear of obstacles and he or she may proceed without delay. Why should the child care that it is a chilly day and leaving the door wide open will lower the house's temperature? Is the child concerned that some unwanted stranger might enter the house? Is the child concerned with leaving the door the way he found it? Will the child stop to think about whether he or she could have opened the door in some other way or speed, which could have eased the opening? The answer to all of these questions is probably negative. It is unnatural for a child to close the door, as this does not help the child immediately achieve the target (reaching the playground and playing). It is even unnatural to think of the optimal way to open doors, as the door is already open and the swing set in the playground is probably more appealing.

As adults, we act somewhat differently. We were taught that the front door should be closed. Therefore, we grew accustomed to closing it and not leaving it wide open (especially on a chilly day!). As we do it daily, closing the door after opening it begins to seem natural, an integral part of the process of going outside. Yet stopping to review the task and consider whether it was performed most effectively, as well as how we should act in future events to ensure that it is efficient, is still quite unnatural to most.

The case with the business debriefing is not much different. Project managers, salespeople, and indeed most of us are busy people. In most cases, the project manager may reach the project's deadline at the last minute, and in many cases, even after the due date. As they approach the project's completion, the managers' thoughts (and hands) are probably set on a new project; maybe their hands are full, completing old tasks that were put aside while working on this project. Stopping to debrief and learn how to best perform future projects may seem reasonable, yet most managers will proceed to the next project rather than stopping and debriefing. Most will assume they lack the time for this additional task.

To make organizational learning possible, debriefing should be defined as an integral part of the work process and not regarded as an additional task. Management should enforce this change. The purpose of debriefing is to develop our knowledge and determine how to better perform future cases based on past experience. Managers on all levels who have a sense of organizational responsibility and wish to succeed should include debriefing as part of their current processes.

Once people grow accustomed to debriefing, it will be as natural as closing the door.

When Should We Debrief?

It is rightly assumed that constant debriefing is impossible. We cannot debrief all day, every day. We must work. Progress must consume the majority of our working time. So, when should we learn? When is it right to halt the work process to proactively analyze our performance and learn how to best perform?

There is no single correct answer. Naturally, every organization must individually decide what triggers its debriefing process. Each organization is unique. Therefore, the question when to debrief and learn should be answered regardless of other organizations' past and current decisions. This is true only in theory, however. In practice, organizations work in similar templates: name an organization that does not handle processes, try finding an (knowledge-based) organization that does not perform projects, imagine an organization that never encounters an unexpected event. These are typical examples of situations that recur in most organizations. Naturally, all of these are situations in which debriefing should be considered. Following is a list of general situations in which debriefing can prove beneficial:

At the end of a project: A project's completion is a good time to stop and review the project's achievements, expenses, and resources. When the project's completion is regarded as a well-defined stage, management can cement the debriefing as part of the project's official life cycle. Furthermore, the debriefing process can be added to the project's organizational work procedure. Many organizations debrief only at the end of projects, as it is an easily defined anchor point. It can be defined as one of the project's stages, making it an integral part of the project process. When the project's duration is longer than usual, it is advisable to debrief at the end of every stage. Another option is to debrief after a milestone is reached.

At the end of core processes: In the U.S. Army, the most popular debriefing technique, the After Action Review (AAR) process, on which I will elaborate in Chapter 4, is performed at the end of daily training sessions. At the end of each training day, 10–30 min are dedicated to debriefing. As a rule, debriefs can and should take place after significant or core processes: training sessions, bids, army operations, and so on. This is not as trivial as it might seem. Working with organizations for many years, I have noticed they tend to debrief significant processes while not debriefing core business processes. Thus, because of their small size and allegedly limited impact, people believe there is no reason to debrief core processes and daily routines. This belief is not shared by the U.S. military. The soldiers debrief regularly at the end of each training day. Training is a daily, normal process. Were they wrong? Were they investing their time unwisely? Quite the contrary. The regular, routine processes are performed more than any other processes. Learning from them is *very* wise, as their impact exceeds the impact of significant, yet rare, processes.

Furthermore, training soldiers do not debrief solely at the end of the training process (usually, 4 days). They debrief at the end of every single day. Debriefing every day enables the crew to improve on a daily basis during the training, without having to wait for the next group, and thus delay the implementation of lessons learned.

Although debriefing usually takes place at the end of projects or processes, a common reason for debriefing is an unexpected event. We naturally debrief after special occasions and events, especially after discovering a malfunction. We experience this phenomenon daily, merely by watching the news. If, for example, an organization lost $2 billion on a deal, it would surely debrief following the loss. If several people were unfortunately killed in a car accident, safety authorities investigate the accident. An investigation as a process can bear much similarity to debriefing. One might even wonder: are these two terms synonymous? They can be synonymous in some cases, or very different in others, depending on the main objective defined by the organization. If the process's goal is learning to perform better on future occasions, we might as well call it debriefing. An investigation may have additional objectives as well—for example, finding those responsible for said failure. In short, if an investigation's purpose is improvement; it may be regarded as a debriefing.

In many cases, although no one forced us to formally debrief, after unexpected occasions, many of us will instinctually begin wondering: *What did we miss? Where did we go wrong? What could we have done differently to prevent the fault, failure, accident, and disaster?* Although such debriefing does yield some results, it is not a debriefing process. Debriefing is performed using organized methodologies and in a group rather than in each individual's head.

Debriefing should be defined as "lessons learning after occasions," with an "occasion" being more than just any occurrence.

It may be a positive occasion or a negative one. To be honest, as much can be learned from positive occasions as can be learned from negative occasions, although it seems easier to learn from negative occasions. It is easier for us to ask what went wrong and then try to analyze those reasons than asking ourselves why everything proceeded as usual. Daniel Levitin, in his book *The Organized Mind* (2014), discusses the issue of memory and remembering. "Cognitive scientists," he writes, "have suggested that we tend to learn more from negative information than from positive... positive information often simply confirms what we already know, whereas negative information reveals to us areas of ignorance" (Levitin, 2014, p. 216).

When surprises and change occur: So when should one debrief? Should one debrief after each occasion? Probably not. One should seek those occasions that are interesting. If everything seems normal, debriefing may be difficult. Let me be more precise. A useful definition as to when to debrief could be to learn lessons from *surprises*, whether good or bad. If an occasion surprised you, you should debrief.

Referring to debriefed occasions might come as a surprise. We are not accustomed to thinking about *surprises* in business-related contexts, and so using *surprise* may seem somewhat stilted. Such a concept, then, requires an explanation. Debriefing on surprises has two main advantages. One is related to the process of debriefing, and the other relates to the outcome.

The process of debriefing on a surprise is easier: we begin by examining the elements of the study that behaved in an unexpected fashion. We examine these elements to understand what should be repeated or avoided. When we debrief on occasions that ended predictably, initiating the learning process is difficult. We fumble for a "starting point": what aspect should we learn from? Where will we find something new? If it was a negative event, for example, the question could be: why did the event occur? This question actually is just as relevant for good surprising events, yet it is seldom used. Using the surprise as a trigger helps us to learn our lessons.

Furthermore, when surprised, we grow curious as to the cause of this surprise. We are motivated to debrief because we naturally wish to solve this mystery. Most of us feel uncomfortable with surprises and attempt to predict the consequences of our action. Our discomfort with dissonance helps us effectively carry out the debriefing process.

As to the outcome, when we are surprised, we know that something must change. If we failed in winning a big deal, we will search for alternative work processes. If we succeeded and were surprised (yes, sometimes we are surprised at our success), we feel lucky. We then try to change the way we handle things to repeat this success on our own merit. Whether we succeeded or failed and were surprised, we know change is required. But if everything went smoothly, debriefing is more complicated. We probably are unsure whether a lesson to be learned exists, thus making us unsure whether debriefing is worthwhile.

We started off this chapter by stating that all organizations handle processes, projects, and events. We then discussed debriefing following the completion of a project (or after reaching milestones), debriefing at the end of (core) processes, and debriefing after surprising situations. Projects, processes, and events are opportunities for debriefing and learning. Most organizations have another common denominator: they all change; some of them change rather frequently. Some of these changes are triggered by the organization's actions (e.g., changing the organizational structure), and some are caused by external factors (market changes, technological changes, and so on).

Debriefing could be beneficial following these changes, especially after the change process has been completed. Paradoxical as it might seem, although the world constantly changes and so do we, planned changes only sometimes succeed. When they do succeed, their results hardly resemble our predictions. In *A Sense of Urgency* (Kotter, 2008), Kotter states that according to his research, most organizations fail in change management during the first (yes, the first!) of the eight stages of change management. Debriefing on these changes suits the criteria to debrief the unexpected. If an organization did not succeed in managing change, one should ask why and should inquire what can be done to ensure success in managing future changes.

If the organization did succeed (which, as explained, is not as trivial as it seems), one should inquire whether the defined targets were achieved and whether the process itself was efficient and relevant to future changes.

Each organization should compose its own list of debriefing questions. An aviation plant I worked with, for example, decided to use this mentioned criteria list and to debrief: following a project's completion; at the end of core processes; when surprises and changes occur; additionally after significant quality failures (customer complaints or in-house failures that caused the organization to violate regulations); and after purchasing new equipment, machinery, and building projects.

Another organization I worked with based its profits on large deals. The organization decided to debrief after a deal is decided on, whether they won the deal or lost it.

Who Should Debrief?

One of the organizations I have worked with discovered that handling debriefing sessions was quite challenging. Workers did not attend meetings, shouting was common, and most concerning, lessons learned were not truly implemented. All this changed once management decided to change the debriefing format and let each group lead and handle their debriefings independently. Before this change, debriefing sessions were handled by an appointed worker. The employee was considered to be an expert on debriefings, accurately analyzing the case's details. Unfortunately, workers felt intimidated by this expert. The debriefing sessions felt to these workers like an investigation, and they feared its results. Because the appointed worker was an expert in the field of debriefing, they could hardly object to the recommendations. And so time after time, people felt their voice was stifled. This eventually caused them to view the debriefing as pointless. Although management tried moderately changing the methodology, most workers resisted cooperation and kept to themselves. This case teaches us that the person conducting the debriefing (the "who") matters as much as its timing (the "when").

The "who" question can be divided into two subquestions: (1) Who should lead such a process? and (2) Who should attend?

Every organization must address this question individually; there is no objectively correct answer. Debriefing is a process that requires methodological understanding as well as objectivity to analyze the issue uninvolved. Lesson learning is not always simple, nor is analyzing the causality of a case trivial.

Yet, as we learned from the previous example, working with an outsider has its disadvantages. These disadvantages stem from two main problems. The first is fear of sharing unpleasant situations with a stranger. The second is the tendency to abdicate responsibility and relinquish it to "the professional" whenever possible. These two challenges relate to one another.

Each organization, based on its specific conditions, should decide whether it prefers using an outsider or insider for debriefing. It is preferable to distribute this task among as many workers as possible.

I recently spoke with two organizations that distribute the debriefing among multiple workers, yet each deals with the challenge differently.

The first, a jailing organization, initially used a hired professional: a debriefing officer who accompanied every debriefing session. After 2 years, they distributed responsibility and knowledge between all organizational units, each unit appointing one of its officers in charge of debriefing. Those officers then were trained and began debriefing in their units, with some assistance at the start and independently later on. The debriefing was subpar. Management reasoned that centralized debriefing may have caused the units to lack a sense of responsibility for the process. They decided to appoint each unit's deputy director as the officer responsible for debriefing. They hoped appointing a high-rank officer would get the job done.

The second organization, an aviation plant, was unsatisfied with debriefings as well. All debriefing sessions were to be handled by the deputy director. Sadly, few debriefing sessions took place because of the deputy's limited availability. Management analyzed the situation and decided to alter their debriefing method. Instead of using a complex, thorough method, they began using AARs. They added two stages to the classic AAR structure: one before the review and one following it, both handled by the deputy director. Now, the core process is handled by low-level managers. This change occurred 6 months before writing this and already is regarded as a success. Dozens of debriefings have taken place since, all led by low-rank managers and all producing high-quality lessons. The quality of these debriefings now surpasses the quality of the deputy director's debriefings.

These two examples clearly demonstrate the challenges organizations face and how various solutions can be implicated according to the needs of each organization.

As mentioned, the "who" question has two subquestions: (1) Who should lead the debriefing process? and (2) Who should attend? The first we have explained; the second is (happily) easier.

More attendants does not mean more efficiency; in fact, quite the contrary. The debriefing process manager should handpick only the most relevant workers, the criteria being their potential contribution to the learning process. Divergence benefits the learning process as it provides us several points of view.

In cases that require many attendants, the debriefing may be divided into subsessions. Results can be integrated later in the process.

How to Debrief?

We have dealt with the "when," as well as the "who," but what about the "how?"

The "how" may differ from organization to organization and even between the units. Yet some guidelines are common to all techniques.

Well, how do you debrief? What must you know?

These questions deserve a chapter of their own.

References

Kotter, J.P. *A Sense of Urgency*. Cambridge, MA: Harvard Business School Publishing, 2008.

Levitin, D.J. *The Organized Mind: Thinking Straight in the Age of Information Overload*. New York: Penguin, 2014.

3
Deepening: Debriefing Techniques and Guidelines

Debriefing is an intuitive process. We all debrief. Take for example any baby with whom you are familiar. Now picture this infant, learning to get up and stand. He will try and err, try and err, and after a few weeks, he will stand. He will probably improve with every try. Why? It might be because his muscles grow stronger with each attempt; but a highly possible explanation is that he is also improving his technique. Here is another example: if someone is repeatedly unpleasant to this baby, the baby will learn to not approach him or her and avoid communicating with this person altogether. We can think of a dozen other scenarios in which people debrief. Even animals debrief (at least Pavlov's famous dog did!).

If debriefing and learning are such natural processes, why dedicate a whole chapter to debriefing techniques and guidelines? The answer to this question lies in life's complications. True, the baby debriefs naturally. But if we would learn to communicate better and teach him or her to debrief more effectively, we might be able to decrease the amount of falling. In organizations, the situation is even more complicated. Too many times, debriefing sessions yield few results. Most workers are preoccupied with more immediate issues, seemingly more urgent than debriefing their own work, or cannot even make time to read the debriefing recommendations of others. In the rare case where we do decide to debrief, we usually do not fully intend to learn as much as possible from the debriefing process. In many cases, we fear a thorough process will require too much time and energy and prove ineffective. Whether ineffective debriefing is caused by lack of motivation, time, or knowledge, techniques and guidelines should be provided. Implementing these can lead the organization to successful debriefing, leading in turn to better results. As the saying goes "there are no free lunches," but optimal debriefing is not necessarily time consuming.

Techniques of Debriefing

After Action Review

Most of us are familiar with only one debriefing technique. People with whom I speak always believe the debriefing technique they know and use is the only one that exists. This is a common misconception. There is and can be more than one valid debriefing technique. In most cases, it is better for each organization to stick to one methodology; however, some debriefing processes or special occasions may require different types of techniques. Debriefing, like many other organizational processes, is a means to an end and not an end in itself. We do not debrief for the sake of debriefing; we debrief to improve performance. The effort we wish to put into the debriefing process is proportional to what we believe can benefit from it. When the price of ignorance is high (i.e., not learning and thus not improving), we probably should invest more time and effort. This can mean using the same technique more thoroughly, adjusting our regular technique, or, in some cases, even using a different technique. I met with a large organization that was dealing with defense products and projects; it defined one procedure for all types of processes and projects with the exception of safety events. These safety events had their own debriefing technique, which was different from the company's general debriefing technique.

To enable management to decide what is best for the organization, getting familiar with some debriefing techniques and understanding their rationale is highly recommended.

Naturally, after action review (AAR) is a good technique with which to begin; it probably is the most popular debriefing technique and is used worldwide. This technique, mentioned briefly in this chapter, addresses four questions:

1. What were our expectations?
2. What happened?
3. How can we explain the unexpected difference between the two?
4. What do we recommend?

These four steps seem simple, and you could say their ingenuity lies in simplicity. Although seemingly simple, these ideas are not necessarily simple to develop. What makes these questions unique is that they are easy to follow. They guide you through a learning process, developing new knowledge based on past experience. The first two questions should be asked as a set, comparing one with the other. Detailed descriptions are unnecessary. They are a waste of time and energy. Look for the unexpected result (as presented in step 2), and search for surprises. Refer to events and results that did not conform to your expectations (as described in step 1). When unsure of the necessity of some facts, they can be added and set aside. The details can be filled in later if something is found to be missing. This avoids including unnecessary information. This method is efficient for several reasons. First, overtly investing time and energy during the "What happened?" step leaves us less energy for the following steps. Second, it distracts our attention from the more important information, on which we should be focusing when delving into the next steps—that is, explaining how these unexpected differences occurred (step 3) and making recommendations (step 4). In some cases, however, people insist on writing down everything that happened instead of focusing on the unexpected. This insistence may be because people find it difficult to drop habits. It also could be due to some workers habitually documenting details in print out of fear of forgetting them. During this process, refrain from arguing and wasting energy on proving peers wrong. Instead of attempting to reeducate those dedicated to summarizing, encourage a thorough writing of an annex and allow only the unexpected facts into the debrief itself. At this stage, it is critical to be concise, as information overflow can kill the process. Some workers may be hesitant to include some information, unsure of its necessity (while still on step 1). This worker may opt to add it, thinking it might be useful later. I recommend the contrary: facts can be omitted. In retrospect, if they are found to be relevant, they always can be retrieved and included. This is not a "one-way street." If while working on step 2, you define something expected that pertains to some "missing" facts, you always can go back and add them. We always will have a second (and third and fourth) chance to complete anything we may have missed. Some people argue this process is not comprehensive. They might prefer to list all the facts and move on to the next phase only after every bit of data has been collected and documented. While I agree that listing all the facts is a more systematic approach, many organizations that prefer the "systematic" method (listing all facts top-down) usually drown in information. As explained, this information flood consumes all resources and does not leave enough time or energy for the next, more important steps—steps 3 and 4—which are the core of the learning process. Several years ago, I was invited to attend a debriefing process, which was to end some stage at a large dairy products plant. I was proudly presented with the debrief: a 160-page report, more than 130 of which consisted of mere lists of facts!

The ordering of the steps utilized in the classic AAR technique may be altered. AAR begins with the expectations (i.e., the goals of this process or event), and only then asks what actually happened. Personally, I find it more convenient, in some cases, to start with "what happened?" Every organization, however, can choose whether to start with "what happened?" and move on to "what was expected?" or vice versa.

Returning to the discussion of technique, step 3 is of key importance. This question focuses on understanding the "why?" Why did something unexpected occur? Why did the project take double the time anticipated? For example, if the answer to the latter is that we did not take some factor into consideration when planning (e.g., the actual time spent for subprocesses X and Y), we re-ask the question: why were these subprocesses mistakenly thought to be shorter? The well-known "five whys" methodology (repeatedly asking why until the root causes are found) can be useful here.

The next step is the core of any debriefing process: reviewing and deciding what should be replicated or avoided in future events. This stage is not without obstacles. As life is complicated, successes (and failures) may have multiple causes. We tend to identify the most noticeable factor as the cause, falsely believing it to be the most significant factor. We must sincerely ask ourselves whether this factor is essential enough to explain the gap between our plans and reality. We must refrain from acquiring information beyond our area of jurisdiction. When numerous colleagues are involved, it is convenient to search for the mistakes of others. Remember: debriefing sessions are not investigations. They are meant for learning, not blaming. If it is not yours to debrief, just stay out of it. Learn your lessons; they will learn theirs.

Another typical problem in this stage is finding the process of pinpointing the main cause of the problem or success to be inconsequential. People tend to jump to conclusions and settle for the first reason that can be perceived to be the cause for the events. Thoroughly searching and finding the root factor is important. This is what makes the whole process of debriefing worthwhile.

When groups find it difficult to understand the root cause, they should add someone external to the process. Two possible alternatives can be considered: adding another employee to the team, or adding a debriefing expert. Although this approach has its disadvantages (extensively explained in this chapter), an external person can more easily notice things. Using an outsider's point of view can become a regular part of debriefing. Using someone internal seems cheaper, yet an outsider may prove more effective and can lead to more substantial learning. You may recruit someone (perhaps from the knowledge management team) who has the skills and experience required to find the needle in the haystack. In this sense, learning to debrief is like drivers' education. Even after passing the test, your driving skills still require improvement. Some drivers improve over time, whereas others remain average. A tutor can help you improve your independent driving skills. If the debriefing expert accompanying the session uses his skills not only to find the root cause of this specific event, but also to teach us how to find it independently as well, most of us can improve our debriefing skills. Eventually, we will be able to manage such a session ourselves, now knowing how to find significant causes.

Moving on to step 4 (what do we recommend?) is usually relatively easy because its answers suggest what we discovered during step 3.

This is the basic structural definition of AAR. One can complete an AAR session by answering only the questions asked in these four steps. Nevertheless, the AAR process can be improved by adding the following framework.

Enlarging the AAR Method

In many cases, the AAR method just reviewed is sufficient with no additional components required. This is especially true when the AAR is informal. This, however, is not always the case. In some cases, the AAR should be regarded as a core process, yet additional subprocesses may be added beforehand as well as after to enable improved learning.

"A Leaders Guide to After-Action Reviews" (1993) is the U.S. Army's guide to AAR. This guide includes explanations for AAR's settings and the four-step process, and it also includes additional pre- and post-steps. These additions turn the AAR core process into an institutionalized learning mechanism. Professional literature presents several examples of all types of additions.

The following example demonstrates an enhanced AAR process, as well as why and when such a process is required. This enhanced process was used in a dozen large Israeli organizations, municipalities, and industrial companies. To understand the necessity of such an expansion in this specific case, understanding the background of this case is vital. Debriefings had taken place at one of the Israeli organizations, an industrial plant, for decades. Yet management was seeking to change their debriefing methodology as they found their current process too demanding and complicated to follow correctly without the help of an expert. By switching their debriefing technique, this organization hoped to enable the manufacturing plant's line managers to debrief independently (it was customary to involve the plant manager's deputy in every debriefing session). The organization initially was introduced to AAR and it subsequently taught the methodology to its line managers, who in turn taught it to the workers. Yet the AAR was not efficient. Even though the new methodology was much simpler, the line managers were not likely to start conducting AARs on their own; the first debriefing session's results were far from perfect. Then the enhanced AAR was introduced. This enhanced methodology created a "frame" for the existing AAR, adding a short session before the AAR process and a closing session following it. Both sessions in this enhanced AAR are handled by a senior manager (in this case, the deputy manager). In the opening session, several decisions are made, the first of which is the decision to initiate the process. Many organizations have established procedures as to when AARs should take place. Yet as AAR is a general debriefing form that has the potential to improve learning, it should and can be used not only when required, but also in many other circumstances in which we believe business can be improved by using this learning method. This opening session takes place even if it is obvious that an AAR is required according to organizational procedures because it has several additional outcomes. At the opening session, we assign the role of AAR leader and choose relevant attendants. The last decision to be made in the opening session is probably the most important one: the senior manager states the guidelines for this debriefing. These guidelines define this debriefing's boundaries, as every process or event can be analyzed from many perspectives. You can, for example, analyze the management of a product development project or its engineering aspects; or you can study the consequences of a product's marketing or delve into the partner's and supplier's aspects. If the AAR leader is not presented with specific guidelines, he probably will analyze the case according to his point of view.

Defining the aspects to be analyzed is a decision that should be mandated by the senior manager who is scoping the debriefing—whether it should be large or focused, depending on how many and which dimensions he or she understands to be crucial, from an organizational management perspective. This decision also defines the people who should take part in the debriefing process, and it allocates the resources to be invested in the process.

Additional decisions can be made during the opening session; every organization has its own needs and nuances. What is important is that the opening session initiates the AAR process and provides directions for the learning process.

This framework includes a final session, again held by the senior manager who initiated the debriefing. During this final session, the assigned debriefing leader reports back to management (at this stage, full attendance of participants is optional). This report includes a list of the team's findings as well as its recommendations (based on step 4). This closing session is not a "replay" of the debriefing process, nor is its purpose to conduct further analysis. This session focuses solely on the recommendations. The senior manager considers whether to accept each recommendation. If accepted, the recommendation's implications are considered. Is there a lesson to be learned here? Or perhaps is a task required to fix the problem and improve?

The difference between tasks and lessons is not an obvious one and calls for an explanation.

A *task* is a duty assigned to a specific individual or group. This assignment must be performed by said individuals at or by a certain time. Every task, therefore, includes three components: *who* is in charge; *what* has to be completed; and *by when* this change should be completed.

A *lesson*, however, is a general recommendation for future acts and processes; it is the best route to take if ever introduced to any situation for which the lesson is relevant. It has no due date, or specific person in charge of completing it.

The difference between these two is more easily illustrated as an example. Let us assume that in a debriefed case, a machine broke down. Let us further assume that one of the debriefing team's recommendations was to conduct a weekly maintenance routine on the machine. Assuming the machine has standard work procedures, management probably will decide to update said procedure and train the maintenance teams. Deadlines for both will be stated and some specific employee will be assigned the responsibility for the fulfillment of these duties. This is a classic example of a *task* derived from a recommendation. If, for example, the root cause highlighted a problem that could occur on several other machines in the same category, additional tasks may be assigned as well (updating the relevant procedures of other machines and relevant trainings).

As an organization, however, a *lesson* can be learned as well. In the future, whenever a new machine is purchased, workers should take precautionary measures when handling these machines. This also may affect the examining process before purchasing new machines, as well as the maintenance procedures written after purchasing. Nevertheless, for the time being, no specific task was assigned to anyone. We know something new that should be considered in the future, yet we cannot assign any concrete tasks to be performed.

Assume also that future recommendations should be embedded in the procedures defining them, and therefore should be converted to tasks that assimilate this new knowledge in the appropriate work procedure. Not every action we perform in the organization is defined by procedures and guidelines, however, because procedures and guidelines consist of defined instructions. A lesson can stay in the form of a recommendation: "When managing an organizational change, it is recommended to communicate the criticality of the current situation to the stakeholders, to explain the need for change, and to perform it as soon as possible."

However, even the most methodical organizations could not possibly specify work procedures for every possible potential process. In many cases, no specific task needs to be assigned, even though a lesson is worth documenting. In some cases, a task is to be derived from the recommendation; in other cases, a lesson should be derived; and in yet some other cases, both are derived, sometimes from the same recommendation—one dealing with fixing the current situation, and the other referring to guidelines for the future.

Deriving tasks and lessons from a recommendation is an important mission that should be executed by a senior manager as in most cases he or she has a wider perspective than the worker leading this specific debrief. This is particularly true given that the senior manager considers more than just what was learned from this specific case. These opening and closing meetings—beginning with assigning a debriefing team and the debriefing guidelines and focus, and concluding with deriving tasks and lessons—create the framework for the actual debriefing.

Concluding with specific tasks and lessons learned is important, no matter which debriefing technique one follows. This step should be implemented as part of any learning to ensure improved future performance.

Multi-Case Learning: Learning from Several Processes or Events

So far, we have introduced one technique: AAR. This technique is more than suitable for a single process, project, or event. In many cases, however, we wish to improve future organizational and professional conduct based on analyzing the results of more than one past project or event. Let us picture, for example, an organization that repeatedly finds itself late on delivering projects. This organization may debrief each event individually. Yet if new projects continue to be delivered late, it might be beneficial to debrief a group of projects. This debrief will focus exclusively on the lateness aspect. I highly recommend this approach to organizations debriefing a phenomenon: instead of debriefing on each case at length, analyze them together; debrief on a set and not on a single event. Unlike AAR, which focuses on a specific event, process, or project and is not necessarily targeted toward a defined business goal, this debriefing session has a clear problem to solve. The rational is rather simple: testing a group rather than an individual is more efficient. The sense of urgency is clearly understood as the trigger to this debriefing is a recurring phenomenon (cf. Kotter, 2008). Conclusions can be drawn successfully as they do not depend on one specific case, but rather take several cases into consideration.

A possibly useful technique in these cases is multi-case learning (MCL). Like AAR, MCL is a simple technique defined by several steps:

0. Define the action requiring improvement.
1. Briefly describe each project, processes, and event.
2. Do these cases have a common denominator? What sets these cases apart from similar events with different results?
3. Can any common attributes explain the results?
4. What do we recommend?

One may wonder whether this technique is an actual debriefing process and not just another learning methodology. *Debriefing* is defined as learning performed after a process or action; reviewing the past to learn how to systematically improve future performances. Given this definition, MCL is indeed a debriefing technique. We review a list of past events, projects, or processes to understand what caused their results, and we recommend future organizational behavior based on our conclusions.

Let us delve deeper and explain how these four questions lead to learning and ultimately enable the organization to improve future performance. Note that defining the action requiring improvement (step 0) is not listed with the other steps. This should not come as a surprise, as this step precedes the actual debriefing process; it triggers the decision to hold an MCL, and includes determining its main leader, focus, and which people and case samples are to be included.

The first step is to shortly describe each sample case (projects, processes, and events) to be analyzed. This step should be performed carefully. Let each case's representative describe the relevant case themselves and not allow others to report on their behalf. Reports given by the representatives enable authentic description and prevent some of the judgmental interpretation that may come from an outsiders' point of view. Authentic reporting will be critical during the next steps.

The cases should be narrated in a "storytelling" fashion, rather than exclusively providing technical facts. Storytelling reminds both the narrator and others familiar with the case of what actually happened. Furthermore, it may help other participants to recall many details (not only those described by the narrator). It is important to keep the stories short, as we do not want to spend much time on this stage; it should be regarded as an introduction—that is a quick recalling of the case's details enabling further brainstorming.

The next two steps deal with identifying patterns. They consist of searching for elements that can explain what has occurred or was viewed that led to the unexpected results. These are obviously not easy steps to perform. In some cases, we may not easily find the pattern. In other cases, we may detect some attributes that superficially may seem to have caused these results, yet we must not exclusively focus on noticeable attributes. We must seek root attributes, as future improvement—not excusing past mistakes—is our motivation. Root attributes are those that may not be noticeable, yet can be deducted through thorough analysis. Although many cases can be handled without discovering the source of their causes, some cases require "digging" for root attributes. As life is complex, most situations result from multiple causes. We should try to find multiple attributes, each common to a subset of the sample cases (although not necessarily to all). We may discover that in each of the cases analyzed, three out of five attributes were found—and this combination, indeed, could explain the results. To validate such a pattern, we must perform some checks. We have to analyze additional sample cases that had different results and then prove that when different circumstances ensue, the same problem does not recur. If this indeed is the case, our findings are validated.

Consider another example (based on a case of a cyber company). Let's assume we are debriefing on projects that have been delivered late. Let us further assume we discover that projects with more than two subcontractors and a young project manager (one with less than 2 years' experience managing projects) are all late on delivery.

We then must stop and ask ourselves several questions:

1. We should ask whether other (additional) sample projects with more than two subcontractors and a young project manager were completed by the project's deadline. If so, we must continue searching for additional attributes that might differentiate the successful projects from the unsatisfactory ones.

2. We should ask another important question that might lead to an alternate finding. It is possible that only one of the multiple attributes is the cause for the reported problem. Maybe working with more than two subcontractors is the only cause for delivering late, making the project manager's age or experience irrelevant.

The MCL's last step (step 4, What do we recommend?) takes us back, as with the AAR, to focusing on the future. What tasks should be performed in future projects according to the patterns found? Our recommendation should be reasonable. Not implementing projects with more than two subcontractors, for instance, is usually an unrealistic recommendation. It might prove more effective to define other recommendations, such as "Although working with several subcontractors can effectively provide a top-notch solution, caution should be taken concerning timelines, as these projects tend to take more time than if performed with fewer subcontractors, and this additional time must be considered when planning timelines."

I doubt the average reader is surprised by the MCL debriefing methodology. It is intuitive and natural, and it reminds us that what we can learn from analyzing a group of events cannot always be learned from individual cases. Nevertheless, most organizations do not consider MCL as part of their debriefing efforts and do not perform it as systematically as they perform AARs.

A more important lesson can be learned here. These two methodologies might compete against one another, yet an organization might decide to change their chosen methodology and not use the AAR or the MCL. This is important because organizational culture can lead to less effective use of these two methods.

Some organizations mainly self-debrief. In these cases, perhaps a third method is required. I witnessed a pharmaceutical organization in which workers came to debriefing sessions well prepared; they all came with personal recommendations (these usually described how their peers' work could improve). A unique methodology was composed to overcome this obstacle, enabling true analysis and learning, rather than debating and assigning improvements to each other. In this case, a double-loop debrief took place. In the first round people were asked to share their understandings and recommendations.

After writing these down, a second round took place where each attendee chose from all recommendations what they thought they could learn to improve themselves and why. Learning was achieved.

No single technique is superior. Every organization can and should choose a methodology that suits its needs and nature. Not only can it choose from the existing methodologies presented in this or other books, but the organization can also develop its own methodology.

Organizations should consider the following two recommendations when selecting or developing a debriefing methodology:

1. Whatever methodology the organization chooses, it should be one that includes truly understanding the case's details and causes.
2. The organization should make sure the chosen methodology focuses on the future and on lessons that can be learned to improve future organizational performance.

Of course, like every methodology, this methodology should be simple and systematic.

We have now set the first building block of lessons management—debriefing processes. Its outcomes are lessons, shown in the below figure:

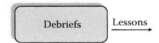

The lessons and practices cycle—debriefing.

Debriefing is a process—a process that enables us to develop new knowledge and learn; however, it is not the only process that serves as a source of new knowledge. Along with debriefing processes, we find two additional processes: quality-based processes and experiencing. These two processes will be described in Chapters 4 and 5.

References

Kotter, J.P. *A Sense of Urgency*. Cambridge, MA: Harvard Business School Publishing, 2008.

U.S. Army. (1993). A Leaders Guide to After-Action Reviews. http://www.au.af.mil/au/awc/awcgate/army/tc_25-20/tc25-20.pdf (accessed February 10, 2013).

4

Learning from Quality-Based Processes

"Cottage cheese bubbling is achieved by adding flour to the recipe. Boiling cheese for too long can give it a doughy texture. This process, therefore, should be handled gently. Room heat also affects the cheese's temperature, and should be taken into consideration as well."

We do not have to wonder about the importance of this lesson. We can deduce, even as strangers to food technology, that it is an important lesson in this specific industry. The question I wish to raise is about the source of this lesson. By examining this lesson, can we be certain it was learned via a debriefing process?

The answer is, certainly not. When a lesson already exists, it is nearly impossible to know whether the lesson was the result of a debriefing session or whether it was learned elsewhere.

Several years ago, I was asked to plan the professional program for an Israeli knowledge management conference. While preparing the various sessions, I sought people in charge of debriefing in their organizations. Naturally, I searched for quality assurance managers. Quality assurance managers regard debriefing as part of the quality control process. Furthermore, quality-based standards, such as ISO (International Organization for Standardization) and CMMI (Capability Maturity Model Integration), demand single-loop and double-loop learning (Argyris and Schön, 1974) integrated into major processes as projects. I spoke with several quality assurance managers, and indeed found someone who I believed could present an appealing case study at the conference. This quality assurance manager was special; he did not resemble any of his colleagues. He served (and still does at the time of this writing) as quality manager VP at a large high-tech, software-based company. But he did not look like your average quality manager (or a high-tech worker at that). He was a gentle, soft-spoken man. When this quiet man *did* speak, listening seemed worthwhile. And so I did. We sat and discussed debriefing. I spoke of lessons as an entity in itself; apart from the debriefing sessions from which lessons can be derived (this idea will be described in detail in Chapter 6).

He listened, paused, and shared a wonderful idea: "Why don't we expand the range of lesson learning and deal, along with the lessons learned from the debriefing sessions, with other quality-based results as well?" he pondered. "As a quality manager, I am in charge of various quality processes. I am in charge, for example, of preparing company subsidiaries for quality management audits on a worldwide scale. When we perform an audit, there are two possible outcomes: corrective actions and preventive actions. Closely examining preventive actions may lead to learning lessons, in addition to the usual task assignment. Root core analysis (RCA) also may generate lessons."

We began to brainstorm. We discovered that many quality-based processes are learning processes, and nearly all learning processes yield lessons.

Needless to say, the lecture was a success, as a totally new idea is usually well received in these circles. It is interesting how many times after comprehending revolutionary ideas, they seem so natural to us, and we are surprised we did not think of them sooner. Many times, newcomers are surprised by the way these ideas, obvious to them, are regarded.

As to our idea: it sounds very natural, and indeed it is.

Let us individually examine quality-based methods and choose typical examples of methods that may yield lessons. Our basic assumption is that once the lesson is created, like the cottage cheese example presented at the beginning of this chapter, we do not care whether these lessons were the results of a debriefing process or any other validated process that took place in the organization.

Learning from the Plan–Do–Check–Act Model (PDCA)

One of the most popular and widely known quality management methods is Plan–Do–Check–Act (PDCA), sometimes referred to as the Deming cycle, circle, or wheel. The PDCA model was initially introduced by Walter Shewhart in 1939 and later was popularized by William Edwards Deming in 1950 at a lecture he delivered in Japan. The PDCA is defined as a cycle. Every process can and should be viewed as a cycle. After *planning* and *doing*, one *checks* what succeeded and what failed, *acts* upon what he or she has learned, and, next time, *plans* based on this learning and *acting*.

The Plan–Do–Check–Act Cycle.

Let us focus on the check stage. Checking is performed by measuring the current results, studying these results, and comparing them to predictions (defined in the planning stage). The differences, if found, are analyzed, and changes, if required, are suggested. These changes usually yield improved acting.

This stage is essentially identical to debriefing: analyzing past occurrences to improve future performance.

Learning from the Define, Measure, Analyze, Improve, and Control Method (DMAIC)

Another well-known method, inspired by PDCA, is the Define, Measure, Analyze, Improve, and Control (DMAIC) method. The DMAIC improvement cycle may be defined as the main tool used when operating Six Sigma projects (Pande et al., 2001). (Six Sigma is a set of tools and techniques developed by Motorola in 1986 and is still widely used today in many industrial sectors). The *measure* and *analyze* stages may be perceived as the equivalent of the PDCA's check stage (divided into two substages). Indeed, the *analyze* stage is one of this cycle's core stages; this stage deals with identifying, validating, and selecting root causes according to which improvements will be suggested. Usually, tests yield many results (i.e., several causes are found). Analysis thus naturally will result in a long list of root causes.

The organization implementing DMAIC must decide on which root causes it wishes to focus. This situation also occurs in debriefing processes; many recommendations can be made based on examining the difference between the predictions and reality. Management then decides, based on the debriefing, which recommendations to implicate to improve effectively.

The *improve* stage is equivalent to the PDCA method's *act* stage. Both stages deal with the events following the debriefing process, ensuring that the organization enacts the recommendations, enabling systematic improvement. The names of the DMAIC stages closely resemble debriefing and learning lessons as they have the same purpose: improvement, enabling us to prevent recurring faults, and repeat success.

In conclusion, both PDCA and DMAIC include a debriefing subprocess as part of their work cycle. Lessons that can be derived from the recommendations should be regarded as lessons learned from other formal and informal debriefing processes.

Learning from Gemba Walks

Gemba, or to be precise, Gembutsu, is a Japanese term. Its literal meaning is *real place*. The Gemba walk is part of the Lean Management philosophy. It is based on sending management to the workshop floor and managing the organization from there rather from their office chairs (Womack, 2011). Some say it is meant to help management understand problems; others regard it as a discovering and improvement opportunity. The process is quite simple: the manager chooses a route that crosses the plant and slowly walks this route. Usually, managers do not plan what to say and where to look; rather they open their eyes and observe, asking questions on their way. When existing problems are known, the Gemba walk serves for better understanding the situation. If used as a routine, it helps explore unexpected performance. It also can be used to understand the causes of a situation, and then decide how to improve it.

Putting aside the strategy of management from the factory floor, what is the core of the Gemba walk process? If we return to our famous debriefing method, the after action review (AAR), we will recall the four questions. What happened? What did we expect? How do we explain the difference between the two? And, finally, what do we recommend? Returning to the Gemba walk, we can identify the same four stages: we know what is expected from our workers; we want to pinpoint the results that differ from our expectations. This is why we go down to the factory floor. We walk, observe, and ask questions. Later on, after detecting unexpected results (referred to in Chapter 3 as "surprises"), we can analyze these findings and submit our recommendations. At this point, we might ask whether AAR and Gemba walk are different. Are they merely synonymous terms referring to the same process? Do we benefit from using two differently named yet seemingly identical techniques, or are we just fooling ourselves, believing that we as managers have a diverse set of tools and a diverse set of influences (eastern and western society)?

In my opinion, they do differ from another. AAR is a general method and the Gemba walk is a specific tool. The Gemba walk is unique in two ways.

The first way is its focus. The Gemba walk asks, "What happened or what is happening?" Some AAR teams investigate to understand what happened. Yet many teams mainly focus on the difference between our expectation and reality. When trying to improve future performance by using the Gemba walk, we also try to understand the current situation; we hope to find a solution through close examination of the state of the plant or organization.

The second aspect is the location in which the Gemba walk is undertaken. The fact that it is executed not in the office, not in a meeting room, but on the plant floor, near the machine and workers, is what makes this methodology special.

Required improvements noticed during Gemba walks may serve as lessons. Gemba walks produce lessons that assist in improving the manufacturing processes.

Learning from Quality Audits

Surely, no book is long enough to include all quality-based techniques, but quality audits cannot be overlooked. Quality audits are the foundation of any quality management system. They are routinely performed by every organization and serve as the basis for nearly every compliance and quality management standard (ISO series; CMMI).

Quality auditing is the process of systematic examinations of some quality systems; audits are an essential management tool for ensuring that the organization performs as expected, following the predefined work procedures.

So, what do quality audits include, and how are they linked to lessons?

Quality audits are audits performed by internal or external personnel or teams that thoroughly examine processes, objectives, and results. In debriefing terms, these audits compare our expectations with reality. This, again, resembles the first two AAR questions: what happened and what did we expect? Quality audits skip, or at least focus less on, explaining the current situation's causes and root causes (step 3); instead, they directly approach the last stage: defining corrective and preventive actions. Corrective actions are, by definition, tasks that help us correct our mistakes. If, for instance, tools were revealed to be missing from the work station, a corrective action could be used to replace the missing tools.

Preventive actions, however, do not deal with the present; they deal with the future. They deal with preventing the recurrence of a discovered fault or incompatibility. A preventive action is a task that, unlike a corrective one, can produce lessons in addition to its tasks: usually, a corrective action requires further analysis if instructions are unclear. Corrective actions are performed to try to understand what caused these surprising results and how they could be prevented in the future. This may result in training people for the existing procedures. It may result in completely changing procedures. Equally important, it may define lessons. We can derive lessons from these actions and define ways to improve future performance. Quality audits, therefore, may produce lessons.

For example, if an audit revealed that some production hall was missing a drill, an appropriate corrective action would be to see that a new drill is supplied. A preventive action, on the other hand, would be to designate noticeable placeholders for each tool, preventing tools from being misplaced. Thus, the derived lesson would be as follows:

"When designing new production halls, significant indications of tools placement will reduce the chances of loss."

This is another example of how all terms, whether quality-based processes (corrective or preventive actions) or lessons learned, are each distinct yet correlated. This enables us to derive one from the other.

Quality-Based Processes

PDCAs, DMAICs, Gemba walks, and quality audits are all different types of quality-based processes that belong to a large group of such processes. Each specific type is in some way distinct, whether by its targets, characteristics, emphasis, or outcomes. Nevertheless, all these processes share some elements, and the new knowledge they produce can become, with some changes and moderation, the new knowledge we are seeking. This knowledge includes recommendations about what to avoid or to seek during the work process, and they include specifications about the circumstances in which this knowledge should be applied. Using our terms: these are pure lessons.

We now have two building blocks:

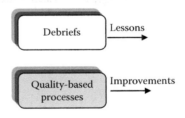

The lessons and practices cycle—quality-based processes.

Conclusion

What should be learned from this is that all types of lessons should be gathered and managed together by one person using a combined system that holds all learned lessons and other suggested improvements. Most organizations appoint one person to handle quality-based processes and another person to manage lessons (if lessons are even managed at all) because they perceive them to be nonrelated entities.

Let us return to the cottage cheese bubbling process presented at the beginning of this chapter. The user could not care less about the origin of this lesson. He or she is more concerned about understanding the lesson than studying the learning process that triggered it. As long as the lesson is validated, and may help him or her to perform systematically better, all other aspects (including the learning method) do not matter. They must be managed, and it seems advisable to manage them together. But, before we describe how to do so, let us examine one more additional source of knowledge: experience.

References

Argyris, C. and Schön, D.A. *Theory in Practice: Increasing Professional Effectiveness.* San Francisco, CA: Jossey-Bass, 1974.

Pande, P.S., Neuman, R.P., and Cavanagh, R.C. *The Six Sigma Way: How GE, Motorola, and Other Top Companies are Honing Their Performance.* New York: McGraw-Hill Professional, 2001, 229.

Shewhart, W.A. *Statistical Method from the Viewpoint of Quality Control.* New York: Dover, 1939.

Womack, J. *Gemba Walks.* Cambridge, MA: Lean Enterprise Institute, Inc., 2011, 348.

5
E XPERIENCE

The *Merriam-Webster Dictionary* defines a lesson as "something learned by study or experience" (2012).

In previous chapters, we elaborated on the "learning by study" part of this definition. We defined two main types of learning: learning by study based on debriefing (Chapters 3 and 4) and learning by study based on performing quality-based processes (Chapter 5). In this chapter, we will explore the second half of the *Merriam-Webster Dictionary's* definition for lessons: "something learned by experience."

Learning from experience sounds rather natural. At some point we have all learned lessons from our experience. It can be the route we prefer to drive home during rush hour; it can be the way a veteran technician knows a machine is nearing its end of life, or the way he knows that it is or is not worth fixing. Another example of lessons learned through experience may be the manager who has interviewed so many candidates for a job he can tell who is telling the truth about their achievements and who is stretching the truth. It can be a mother knowing that when her teenaged daughter returns from school, it is better not to speak with her immediately and let her calm down before trying to understand what happened. Each and every one of us can probably recall dozens of things that we just know, things that we did not learn from anyone else, but rather picked up along the way. These things can include both actions recommended for certain situations (e.g., when it is best to use a specific technique for a new sales campaign) and discouraged actions we should avoid (e.g., in what circumstances we should use restraint and not answer when a customer complains). We call this type of knowledge "experience." Some experience is based on actions we performed that yielded successful results, and therefore we learned how to best handle similar situations in the future; some are based on mistakes we made in the past that we wish to avoid in the future. Some of these nuggets of experience force us to reconsider our methods, and others deal with the way we may perceive things and understand them (Cell, 1984). We learned these pieces of knowledge without any formal or informal debriefing process.

Even without a proactive debriefing session, debriefing processes do take place in our minds. We do not classify these learning processes as debriefing because we do not intentionally stop to debrief. In many cases, the learning happened without us even being aware of the process. With each project, we know how to manage our resources better; with each lecture, we improve our public speaking skills; and with each occasion, we are better equipped to handle surprises.

As a rule, most activities can be defined with regard to these two extremes. If debriefing sessions may be regarded as a type of "explicit learning from experiences," then learning from experiences without intentionally analyzing an event's details (as well as its causes and implications) should be defined as "tacit learning from experiences." The spectrum between these two extremes (deliberately learned knowledge and unintentionally acquired knowledge) is wide, and it fully represents the broad range of lesson learning methodologies. *Merriam-Webster*'s definition of lessons regards these two types of lessons (learned from either experience or study) as separate entities; yet in life, we learn in a variety of ways, starting with explicit study and ending with things we know based on our experience.

What is the process we go through when learning from experience? The following three approaches attempt to answer this question. These theories differ from one another, yet they are complementary. They were defined by three thought leaders: David Kolb, a professor in the Department of Organizational Behavior at Harvard University; David Kahneman, a recipient of the Nobel Memorial Prize in economic sciences for his research on decision-making, professor of psychology and public affairs; and Edward Cell, a professor of philosophy at Sangamon State University. Understanding these three approaches enables us to better understand the essence of learning from experience.

The Process of Experiential Learning: Kolb's Four Forms of Learning

Kolb, in his book *Experiential Learning: Experience as the Source of Learning and Development* (1984), describes the characteristics of experiential learning, revealing two structural dimensions that underlie the process of experiential learning.

Kolb refers to learning as a continuous process grounded in experience; it is best performed when referred to as a process rather than defining outcomes to be achieved, and it involves transactions between the individual and environment. Learning is defined as a process of creating knowledge. He defines the two dimensions through which learning styles take place.

The first dimension is the active-passive dimension: people can learn via *active experimentation* (action learning) or they can learn through *reflective observation*. Some of us find it easier to learn by doing (the active experimentation). We feel we have to experience things ourselves to fully understand and learn. Some of us prefer to observe someone else's actions and reactions, maybe listening to a case study and what that person has learned. When these people hear or read a case study describing full implementation of an idea, they can visualize it, understand it, and learn from it. These two methods (active and reflective) are different channels through which a transformation can occur in our minds, and they both result in learning.

Yet different types of learning can take place through another dimension: the concept-details dimension. Kolb refers to this dimension as the "prehension of knowledge" (1984, pp. 41–42); that is, how people take hold of the knowledge. Some people find it easy to learn through *abstract conceptualization* (grasping via comprehension) and others prefer learning through *concrete experience* (grasping via apprehension). Some people find it easier to understand when a topic is explained top-down (e.g., explained from the abstract level down to the concrete details); some people understand better when new knowledge is transferred to them by describing its implication (understanding in a bottom-up style).

Together, these two dimensions define four styles of possible learning:

- Concrete experience (which is derived from the experience's details)
- Abstract conceptualization (which is derived from concepts we come across)
- Reflective observation (which is achieved by listening to others)
- Active experimentation (which is achieved by doing)

Learning can take place when we are exposed to one or more of these styles. Different individuals will find it easier to learn though different styles, and each person will tend to better understand and learn through a different combination of these styles.

The Process of Experiential Learning: Kahneman and the Two Systems Theory

Kolb teaches us that different environments and conditions can trigger learning (i.e., details, concepts, listening, and doing), but this does not explain what happens in our heads during the learning process itself. Do we always learn? What causes the learning? David Kahneman explains the processes our mind performs. In his book *Thinking, Fast and Slow* (2011), Kahneman depicts our brain as a machine. "Our brain may analyze our experience using one out of two mechanisms, named "System 1" and "System 2" (these terms were first coined by Stanovich and West, 2000)" (Kahneman, 2011, p. 48).

System 1 is what we might refer to as our "automatic mode." If for example, we are asked how much is 2 times 3 we automatically answer 6. We do not start calculating in our heads. We will spit out the answer with no hesitation. But what if we are asked how much is 17 times 24? Most of us probably do not have a prepared answer, and so will start calculating (10 times 24 is 240, 7 times 24 is…). When calculations are complicated (as in this case), people tend to divide this calculation into two simpler parts. Most people, after a minute or two, will reach the correct answer (which, by the way, is 4080). My point, however, is not to prove that it takes more time to come up with an answer when faced with a complicated problem. That is obvious. My point refers to the systems in our brain that participate in solving the problem. In the first calculation, in which we knew the answer without performing any calculation, System 1 was activated. System 1 operates with zero or minimal effort (Kahneman, 2011). Using the terms presented in this book and chapter, no *learning processes* occur in this instance. System 2, however, is different. It not only is in charge of calculating numbers but also takes control when System 1 fails to decide. If a familiar pattern is not identified (something we already know how to respond to), System 2 is in charge of coming up with the answer. If some decision our brain is making requires attention, System 2 begins operating. With regard to learning from experience, it is interesting that System 2 not only operates decision-making, but also may change the way System 1 performs in future situations. The way System 2 reacts to situations (problems, dilemmas, etc.) affects the definition of the patterns System 1 attempts to identify and use. For example, let us assume we were asked to design a new book cover. Later, when the book is published, we discover that the final product hardly resembles our design. Looking at the cover, we realize that changes in the weight of the paper affected the cover's width and therefore affected both the size of the front cover picture and its composition. The system that looked at the new book cover and was surprised is System 1. It then transferred control to System 2, which in turn analyzed the situation and learned from it. System 2 then "programs" System 1: in the future, whenever we are asked to design a new cover, we "automatically" should ask what the planned weight of the paper is. Kahneman teaches us that although we sometimes learn from experience, this learning is activated only when System 2 is activated.

Thus far, we have discussed two approaches: Kolb's approach and Kahneman's approach. We now will explore a third approach, developed by Cell, approaching this understanding of the learning process from yet another angle.

The Process of Experiential Learning: Cell's Four Levels of Experiential Learning

In *Learning to Learn from Experience* (Cell, 1984), Edward Cell describes the learning process from a different angle. Cell opens with the following statement: "most of the time, we are not really learning. We are simply responding to a familiar situation in a way we have previously learned" (1984, p. 41). What Kahneman calls "System 1," Cell names "responding."

How come our brain does not learn from every transaction? The answer is simple: it is too expensive. Learning requires time and effort. Our brain is trained to reuse the existing knowledge and information. Only when the existing patterns are found insufficient does our brain initiate a learning process to decide how to react.

Cell refers to four forms of learning that may take place, explaining what level of learning we can achieve from each.

The first, and simplest, form of experiential learning is *responsive learning*. When learning this way, we change our response to a known situation. If until now we have reacted to a situation with a series of responses, we now add additional responses to the current "response list" or substitute part of the existing list with these new responses. For example, every time a project was running late and schedules were delayed, we updated the customer. If the last time this occurred our boss was furious with us for not checking to see whether we could use external resources to stay on track with the project's schedules, then we do not need to formally debrief. We instinctively understand that from now on, we have two possible alternatives that can be examined and used: (1) delaying and updating, or (2) using external resources.

Some of us might assume that this is the only form of learning, as most of our best practices (learned from experience) and most of our lessons (learned from debriefing) are learned this way. Nevertheless, Cell teaches us that we may reach additional, higher levels of learning.

The second form of experiential learning is *situational learning*. When learning this way, we change the way we interpret a certain kind of a situation. Like responsive learning, situational learning also may result in changing our reaction; however, the difference between these two forms of learning is fundamental. When acting on responsive learning, the situation we face is well understood. What we learn in that case is how to better *act* upon a given situation. When acting on situational learning, we *understand the situation* in a different light. For example, we as parents may be angry with our children for not finishing their chores. Naturally, we will think in educational terms. We will focus on the chores and the importance of performing duties as an equal part of the family. Situational learning can take place if we decide to view the situation differently. When we yell at our child, he might be so frightened that he will not listen to anything we say. He only hears and experiences the yelling and his accompanying fear. Situational learning helps us understand the child's perspective and view yelling at him not as an educational situation, but rather as a frightening one. Hopefully, this situational learning will help us find more effective educational methodologies and ultimately improve our methods.

I must admit, even though I personally have dealt with learning from experience for more than a decade, I never stopped to contemplate the importance of such learning. We never included any learning of this form in a "best practices" knowledgebase, nor did we ever encounter lessons derived from debriefing resembling such learning. Although I have been familiar with Cell's theory for several years, I hardly applied this knowledge. Only now, when writing this chapter, do I truly understand how we can leverage existing personal knowledge and aim for additional lessons, making it all part of our organizational learning.

The last two forms are less relevant to our discussion, but I will share them, so that I do not leave the reader curious.

As mentioned, switching your interpretation of a situation is named situational learning. As simple as this might seem, it is not a trivial task. Learning the skill of observing situations and interpreting them in diverse ways is a higher level of learning. This is our third form of learning, named *trans-situational learning*.

The fourth and highest form of learning deals with altering our concepts or creating new ones. Developing this ability is referred to as *transcendent learning*. Of course, this level requires one to have acquired situational learning also, as knowing how to interpret situations according to various perspectives is a basic step in constructing a new concept.

We ignore these two last levels of learning, but not because they are unimportant; on the contrary, they are extremely important. In this book, we are more focused on creating and using knowledge, and are less focused on creating personal skills that will enable this learning to take place.

Integrating What We Learned

We can now integrate this chapter's sections into one clear conclusion. Learning from experience can be explained through various aspects: it can be explained (1) by understanding the causality of our learning process (Kolb's styles of experiencing and learning); (2) by analyzing the brain's systems that function when we experience (Kahneman); or (3) by understanding the different levels of insight and knowledge that yield from this learning (as defined by Cell). How are these experiences captured? Usually in the most basic way, as Carla O'Dell and Cindy Hubert suggest in their book *The New Edge in Knowledge* (2011): "capture the critical knowledge from a shifting workforce" (p. 161). It is recommended to define times for speaking with experienced people, interviewing them, and capturing their critical knowledge.

Integration, however, takes place on another level—connecting the subjects of Chapters 1 through 4: lessons, quality-based processes, and experience. When combining these, our diagram is widened:

The lessons and practices cycle—experiencing.

Just think about it. We learn as individuals and as organizations. A major part of our learned knowledge probably is based on learning from experience; usually, debriefing plays quite a minor role in our learning experience.

Nevertheless, when referring to organizational processes, we treat these two sources differently. For some odd reason, we cherish lessons that have been learned as a result of studying (i.e., a debriefing process). Experiences are handled rather differently and are regarded as less important.

Obviously, no one prevents people from working better based on their professional or personal experience. On the contrary, organizations encourage each one of us to perform best based on all possible information and knowledge. Yet organizations do not handle knowledge learned from experience in the same manner that they treat explicitly learned lessons. Most organizations do not handle these experiences at all, and those that do handle these experiences, handle them as an independent entity, set apart from other knowledge.

In short, it is rare to find an organization that merges knowledge acquired by experience with lessons learned from debriefing sessions.

So, where exactly *is* experience handled, in those few organizations that do not ignore it? The answer is rather simple: organizational experience usually is managed under the title of "best (or good) practices."

Let us stop for a minute and consider this. Many of us may be familiar with the concept of a "best practice" knowledgebase. This title may have two meanings. Some organizations, when using this term, mean that they hold a library of case studies, including business stories from which lessons can be learned. Other organizations, however, maintain an explicit list of items, each including a best practice line.

Chapter 4 opened with a sample lesson: "Cottage cheese bubbling is achieved by adding flour to the recipe. Boiling cheese for too long can give it a doughy texture. This process, therefore, should be handled gently. Room heat also affects the cheese's temperature and should be taken into consideration as well."

Now, let us ask ourselves again: does it really matter how this knowledge was acquired, as long as we know we can rely on this information? What, if any, is the relation between the origin of this lesson and its benefit to business?

Most people hate debriefing, with the exception of debriefing the work of others. Yet even this is a risky move because you can never predict whether someone will say something that makes you look bad in the after action review. Even though debriefing is not meant for blaming, debriefing activities easily can lead to blame. If the activities we debriefed had resulted in some sort of malfunction, searching for the worker responsible for this mistake would be an obvious follow-up step.

In some rarely found organizations, debriefing is an integral part of the organizational culture, embedded into the organizational DNA. In these organizations, debriefing is not considered a threat to workers. These organizations, however, are a phenomenon. Also, when organizations hold workers responsible for their failures, it surely stunts advancement. People avoid debriefing sessions as they are concerned their failures will be exposed or discussed, and they are afraid these failures will be used against them, either in the near future or sometime in the future.

In addition to the workers' concerns, or even fears, debriefing has other organizational costs. Debriefing is a time-consuming activity intended to take place in a rushed environment. We are always in a rush. We always agree that debriefing is a good idea in theory, but not now. This procrastination ultimately leads to deserting debriefing altogether.

And yet some organizations *do* debrief. Not every organization, and not in all appropriate occasions, but debriefing does take place.

Leveraging knowledge learned from experience is even less popular. As we have explained in this chapter, this is something that must be changed. There is gold out there, and it is a pity not to use it.

Where do we stand? By now, we have three building blocks: debriefing, quality-based processes, and experience. These are all sources of new knowledge.

How should this knowledge be handled, once we collect all of its pieces? This will be explored in the next chapters, dealing with further managing the created knowledge.

References

Cell, E. *Learning to Learn from Experience*. New York: State University of New York, 1984.

Dictionary, M. W. (2012). *An Encyclopædia Britannica Company*, 2015. http://www.merriam-webster.com.

Kahneman, D. *Thinking, Fast and Slow*. New York: Farrar, Straus and Giroux, 2011.

Kolb, D.A. *Experiential Learning: Experience as the Source of Learning and Development*. Englewood Cliffs, NJ: Prentice Hall, 1984.

O'Dell, C. and Hubert, C. *The New Edge in Knowledge: How Knowledge Management Is Changing the Way We Do Business*. Hoboken, NJ: John Wiley & Sons, 2011.

PART III
Managing Created Knowledge

6
ACTIONS

A famous joke describes the disorder of the early years of the State of Israel. It involves two people who are working at a forest clearing. One worker is digging holes, and the other is pouring dirt into the same holes. A bystander asks these two peculiar workers what they are doing.

"Oh," explains one of them, "We are working very hard and planting trees."

"Where are the trees?" questions the passing guy.

"Well," explains the first worker, "we usually work as a trio: one of us digs a hole, the second plants the tree, and the third fills in the hole. Today, our planter is sick."

Chapters 2 through 5 of this book have dealt with generating new knowledge in organizations, whether by debriefing; by running quality-based processes; or merely by working, living, and experiencing.

It is well understood, however, that acquiring knowledge without acting on it is like digging holes and closing them without planting any tree; it is a waste of time and effort.

Most organizations take action as result of debriefing; they define tasks to be accomplished. The defined tasks consist of three components: what must be done, who is charge of doing it, and by when it should be completed.

Consider the following example of such a task that occurred in January 2016: a debriefing process took place after a downtime had occurred at a server farm. This debriefing led to the understanding that holding specific extra electrical equipment could dramatically reduce the duration of future downtimes. The organization, indeed, chose to purchase and install said equipment and was glad that it was not too expensive. The systems infrastructure manager was assigned to this task and was instructed to complete it within 15 days.

Let us look at another example that occurred in December 2001: a child was injured by a coat hanger's facility at his elementary school and nearly lost his eye. The department of education debriefed and instructed all principals of kindergartens and schools that had a similar facility to replace it with one that is safer for children. The principals were instructed to do so within 60 days.

I could go on and on with various examples, but the idea is clear. Some changes can be defined simply and accomplished within a defined period of time. We then can manage these tasks and track them to verify that they actually are being completed as defined.

Task management is part of every manager's toolbox. It has to do with all aspects of an organization's operation, not just with debriefing. It therefore will not be further elaborated on in this chapter.

Action, however, after generating new lessons is much more than just task management. It would be easier if we could manage all outcomes as tasks, yet life is not that simple. Consider an example from school life. Let us assume that the problem was different: a child was forgotten in class at the end of the day as he went to pack up and was locked in the room by the hurrying teacher. This might result in a new procedure stating that teachers had to check all children in on an attendance form as they arrive in the morning and must check them out on the attendance form when they leave. The teacher is instructed not to leave the room unless both steps are performed.

Handling this instruction as a task may be insufficient. Updating the procedure and sending out a message via e-mail may be good ideas, but they are not enough. People tend not to read every e-mail. People tend to forget. People tend to stick to their current habits. If the change is important enough, we cannot settle with only a procedure update and a note. We probably have to speak with the school administrators and guide them to examine teachers' conduct in classes they review. We may want to include a speech by high-ranked personnel in future seminars, explaining the importance of the new procedure and the potential danger of neglecting it. We might decide to include the procedure in safety trainings for schools. In six words: we have to manage this change.

Change management is not a newly coined term. Its roots can be traced back to the middle of the twentieth century, when theoreticians such as Kurt Lewin (1947), Everett Rogers ([1962] 2010), and others developed theories and methods for handling change and objections to change in the organization. The field of change management has been researched and taught theoretically for many years (e.g., see Cameron and Green, 2004), yet it has been put into practice less often, perhaps because of the theoretical nature of the suggested methods.

Throughout the years, additional methodologies were developed, with each decade featuring a more practical method. Modern organizations understand the need for change management. Organizations recognize that in too many cases, changing is not just allocating, performing, and completing a task. Organizations understand that changing is a process, or maybe even a journey. No definitive rules ensure the success of this journey, yet many good methods and good practices exist. (Chapter 12 is dedicated to change management, so we will not elaborate here.)

The bottom line is this: if we want our lessons to yield better operations and improved performance, we must transform the knowledge into action. If the action is straightforward, well defined, and easy to apply, we can assign tasks. If not, we are required to define the changes, and then manage these changes as a process or even a project.

Before I continue, here is a short recap: up to this point in the book we have learned that there are several sources for knowledge (see Chapters 2 through 5). We also have defined the next step toward business improvement. We defined actions to be performed and we differentiated between tasks and changes. Figure 6.1 illustrates this process.

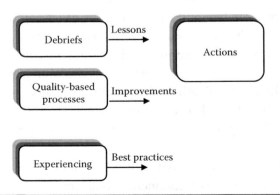

Figure 6.1 The lessons and practices cycle—actions.

This, however, is not the full picture, as one can assume from quickly examining the diagram. Some knowledge cannot turn into action. We have obtained new knowledge that will have some importance in the future, yet it currently is not actionable. How do we remember it for later? How do we retain this knowledge even though there is no change to implement here and now?

We could preserve the knowledge as a written work procedure for future reference, whether as a new procedure or as an addendum. We argue, however, that not all knowledge best fits as part of a procedure.

For example, we might not have any procedure at all: the lessons learned might have to do with how to better prepare for customer visits. Assume no procedure exists for this process (and rightly so because not everything people do can be handled according to procedures).

Another scenario, for example, could be that the organization wants to "loosen" its procedures. I have come across this phenomenon several times when dealing with banking organizations and finance industries. For many years, we have believed two aspects should be manifested in procedures: (1) regulations, defining how we are expected to act, and (2) best practices of the organization and the industry, defining how we would want our organization to act. It sounds terrific and makes sense when adding in these newly obtained lessons. The only problem is that because of the regulatory aspect of the process, we find ourselves controlling (maybe even over-controlling) the procedures to strictly ensure that all employees act exactly as defined. This could be a wonderful tactic for ensuring that regulations are indeed followed, but it is less than optimal for best practices. Although we want people to act upon them, we want to leave some flexibility regarding the exact application of this knowledge. Surely we do not wish external auditors to punish or even fine us when they naturally find that not all employees worked exactly according to these procedures. This situation generated a new trend in which organizations (especially regulated organizations) minimized their formal instructions and work systems procedures, instead of including more regulation and fewer best practices in their procedure.

Consider yet another scenario: we have lessons and good practices that merely serve as recommendations, not as strict instructions. Organizations hesitate whether to include them in the procedures, wishing to provide the employee with a clear differentiation between actions that are obligatory and those that merely are recommended actions.

Last but not least, we have another group of lessons. The lessons that could have turned into changes yet the organization, overwhelmed with other changes it may be required to implement, decided to prioritize only a fraction of them, leaving out all the other good recommendations and ideas. These "leftovers" are wonderful dos and don'ts. The organization must find some way to include these ideas in the organizational memory even though they will not turn into prioritized changes here and now.

How, you might ask, can we solve these dilemmas?

The next chapter suggests a possible solution: storing and handling this precious knowledge in a designated knowledgebase.

References

Cameron, E. and Green, M. *Making Sense of Change Management*. London: Kogan Page Limited, 2004.

Lewin, K. Frontiers in group dynamics II: Channels of group life; social planning and action research. *Human Relations* 1(2): 143–153, 1947.

Rogers, E.M. *Diffusion of Innovations*. New York: The Free Press, Simon and Schuster Inc., (1962) 2010.

7
Knowledge and a Knowledgebase to Handle It

In December 2012, Google was listed as the most visited website in the world. Most of us probably are not surprised to read this. We all are users of Google. Google performs a variety of services, but its search is still (at the time of writing these lines) the most popular service. We all search using Google's search engine, some of us more than once a day. Using Google, or any other search engine, to find answers should not be so obvious to us, however, and such a phenomenon is worth further investigation. Google's search engine does not own any information. We do not visit their website to learn from their possessed knowledge. Google is a mediator, and a great one, too. Google search is so popular that the word "Google" is often used as a verb, meaning to "search" ("to Google it"). If Google is merely a mediator, how is this incredible popularity possible? How can we explain Google's strong influence on our lives? Why do we cherish the mediator itself?

This seemingly unnatural phenomenon can be explained: Google has conquered the field for its perfect ability to deliver the most relevant information answering our needs, even when we define them quite roughly. Before Google's domination, other search engines, such as Yahoo! And Lycos were available. Yet Google succeeded in captivating us more than any other search engine. Furthermore, it changed our habits. It enabled us to get closer to information than we ever could have dreamed. The lesson of Google is loud and clear: accessibility is critical.

As mentioned in Chapter 2, we all experience the challenges presented by information overflow. We must remind ourselves that too much information is almost as bad as no information, as the result is similar: lack of use. If we cannot find what we need because we cannot sort through all of the information and locate what we need, then we will not try and seek it in the first place. We obviously will not use it. When dealing with information overflow, some of us will work hard to locate what is required. Sadly, others will try to locate what they need, but if the required information is not found within a few minutes or even seconds, they simply stop searching. If this recurs too many times, people stop searching altogether. Why bother looking for something when it is simply inaccessible? This assumption applies to almost every type of data, information, and knowledge. Unfortunately, it is the situation for organizational lessons, good experiences, and best practices, as well. Even though people do not want to repeat errors, and even though they want to shorten processes and repeat successes, time is always a scarce resource. People will invest only a limited portion of their time to search for lessons learned while hoping to learn how to perform better. Let us not confuse the issue; this phenomenon is not a manifestation of laziness. We all try to manage our time, and most of us doubt that further searching for lessons learned will lead to something that really can help us improve. This is a classic case of being oblivious to our ignorance. The result is that we may give up on searching in the first place because we are unsure of its effectiveness.

Regrettably, searching for lessons to learn has further organizational implications. Developing knowledge in the first place—that is, debriefing—costs us a great deal of time and organizational or personal exertion. It sure seems to be a waste to invest this time and energy if the knowledge is buried in a huge pile of organizational documents, with no one knowing where or what it contains. Debriefing has slowly faded away in too many organizations because these organizations have determined that the effort is not worthwhile, given that most of the lessons learned have no practical application.

We need to access lessons learned before we perform new activities, to ensure that we systematically succeed in preventing recurring mistakes; avoiding the unnecessary repetition of work processes; and, if possible, enabling the organization to repeat past successes. Theoretically, this does not seem to be a complicated task; in practice, however, this task is far from simple.

KNOWLEDGE AND A KNOWLEDGEBASE

Visualize yourself working for a large organization, searching for lessons learned. In most cases, this is not a thrilling experience. I have had some experience with trying to locate lessons learned in some organizations. I have learned that in some cases, the information is not gathered in one place. Needless to say, it is nearly impossible to learn from past lessons when they are scattered all over. If you do not know what lesson you are looking for, you have no possible way to even begin searching. Furthermore, the lessons that we already know about are all too often the unimportant ones.

But let us not be so pessimistic. In some cases, organizations do manage their lessons. The most common way organizations share their lessons is by collecting and saving all debriefing documents in a shared library. In some cases, lessons are saved in a document management system or portal. Although this is indeed better than not sharing this information at all, it still is not ideal.

Let us picture a situation in which a project manager is starting a new project and wishes to learn from past lessons. Let us even assume that she knows the lessons are stored in the library, knows where it is located, and has access to it. The list of debriefing documents may be long. She has to decide which debriefing sessions she should select to begin her search for the required lessons. Some documents are organized as forms, others as reports or presentations. Our project manager has to carefully analyze each type of file and find where in the document the lessons are located, whether by completing a form or other type of document. Then she has to try to understand the context in which the lessons were learned, and debate whether the examined file is relevant to the new case. If the organization has conducted a substantial amount of debriefing sessions (without mentioning other quality processes and experiencing knowledge, as described in Chapters 5 and 6), this mission is far from simple. The list of files may be rather long; the full task may seem endless.

Normally, the user will select files that bear names that somehow are associated with the current situation. When working with a document management system or a portal, the documents might be organized according to a value or attribute. The user can be more sophisticated. She probably will choose files that contain attributes and values similar to the context in interest. For example, if a project manager is designing a small-scale, high-risk project involving technology and services, she probably will seek lessons dealing with technology and services embedded together, possibly also filtering to small projects and high risk. Focusing on files with attributes and values similar to those of our new case may ease the search process. Nevertheless, this strategy is far from the best process. This heuristic seems natural, and many of us probably have worked in such a manner in some similar case, so let me explain why it is not suitable.

When we analyze the past and develop a new lesson based on it, the lesson is surely affected by the context in which it was learned. Lessons, as defined earlier, are recommendations based on past experience. Therefore, searching for previous cases that resemble the current case could be a good start. Yet life is more complicated. A new lesson developed may be relevant to many cases, not only those of a context similar to our own. We may come up with a lesson after finishing some small project handled in the marketing department, with minimal risk level. Yet the lessons learned also may apply to all high-risk projects, or every project handled in all departments of that organization (not only in the marketing department); it may be an important lesson relevant to large-size projects, as long as they are high-risk projects. When we learn and develop a new lesson, we always start with a specific context (the context in which the lesson was learned), yet the lesson learned usually is applicable to a variety of contexts. A lesson learned from a process of handling audits, for example, also may be relevant to a process of handling customer satisfaction reviews. Some lesson learned by an engineering department may pertain to working with partners and therefore would be relevant to any department in the organization that also worked closely with partners.

There are, however, complications. A single file describing a debriefing session normally will include several lessons. As explained, each lesson may have a relevant set of contexts. Returning to our earlier problem, the file may be tagged with only one set of attributes and values. No combined sub-list of attributes and values will suit. If, for example, we tag a file combining all possible listed contexts of the lessons included, we might mislead and even confuse the user, giving them good reason to ignore these attributes altogether. Naturally, a summary of a debriefing session will be tagged for attributes and values that relate exclusively to the context in which the lessons were developed. This is somewhat useless as every lesson may have relevant attributes and values, defining when this specific lesson would be applicable. If the lessons are not tagged with their specific attributes and values, defining their relevant contexts, how will users know when it is worthwhile to access and learn from them? And if people do not access the lessons and implement the recommendations included in these lessons, what is the point of creating them in the first place?

Thus, even if the organization wishes and attempts to organize the debriefing sessions to make them applicable for future needs, they likely will find this a nearly impossible task. Every file containing more than a single lesson will contain lessons relevant to different contexts, and therefore they will suit different situations. Organizations that wish to be organized and manage their debriefing sessions in some library, portal, or document management system will not find a solution using these technologies, as they are all file oriented and do not enable access for users searching for relevant existing lessons that can fulfill their knowledge needs.

The good news is that a relatively simple solution that addresses this challenge does actually exist. The solution is to store lessons, knowledge, and best practices in what we call a *knowledgebase*.

Saving the Lessons in a Knowledgebase

A knowledgebase is a database containing knowledge. The knowledgebase is an effective way to enable organizations to properly manage their lessons. One might wonder whether there are any differences between a knowledgebase and the libraries, document management systems, or portals mentioned in the previous section. There is only one difference, but it is a substantial one. The latter manage files (various types of documents) while a knowledgebase (in its exact definition and in the context used here) handles the lessons themselves. Each lesson is managed as an independent object and may be viewed as a record. Each lesson is accessible, as its core is a single sentence—the recommendation (what should be performed or avoided). Yet the most important difference between a knowledgebase and other knowledge-management methods lies in the fact that every lesson, stored in a database, has its own set of attributes and values, not necessarily identical to those of the source from which it was originated. Two lessons developed in the same debriefing will not necessarily share the same context defining their relevance, even though both were derived from the same occasion.

For example, let us assume we held a debriefing session regarding a sales opportunity we missed. If we managed these lessons by storing them in a library, portal, or document management system, as described, then we would have produced a new file describing the debriefing meeting and all that occurred in it, including all of the lessons learned in this session. If, for example, we had identified three lessons, they all would be stored in the same file. This file then would be tagged and assigned values. We probably would set the attributes and values to the sector of the potential missed sale, and perhaps to the type of project or technology.

That is the usual scenario. What is suggested here is quite different. The depictions of debriefing meetings are saved in some library, yet they are not the main focus, and instead are regarded as supporting information only. Our three new lessons are inserted into a knowledgebase, each as a separate record, each assigned with its corresponding values. The first lesson may be applicable for all large projects, no matter what type and what sector the organization is dealing with; the second lesson may be applicable only for the homeland security sector; and the last lesson may be relevant for medium- and large-scale projects also defined as high-risk projects.

Saving each lesson as a separate object has two main benefits.

The first benefit is efficiency. The person seeking known lessons does not have to read the full description of the meeting. He can read only the lessons, including the conclusions implying recommendations on what to perform or what to avoid in some predefined circumstances. We all live and work in an environment in which time is a scarce resource; many of us do not have the time or the patience to read the full description of the meeting. Nevertheless, the full debriefing summary document will not be omitted. In some cases, the full documentation can be useful, helping someone who reads the lesson to review and understand how and why the lessons were learned. This option can be provided easily by including a hyperlink from the lesson, linking it to the file in which the summary of the debriefing meeting is stored. We will elaborate on this hyperlink later in this chapter.

The second benefit of storing each lesson separately is the improvement in the accuracy of the lessons. Assume a salesperson is working on a new opportunity in the finance sector. We want our salesperson to be exposed to all relevant lessons learned from past cases (including the case we described in the previous example) even though this new opportunity is set in a different sector than the previous one. Saving each lesson in a separate record enables us to accurately define specific situations in which this lesson is relevant. It also enables users to access any specific lesson that is applicable, even if the case on which these users currently are working has a different context. By dividing the lessons into individual records, we obtain accuracy because every lesson is tagged with its specific appropriate attributes and values, thus defining the context most relevant to its implication.

Both efficiency and accuracy deal with improving and easing accessibility. Both internal and external accessibility require improvement: improving internal accessibility eases the process of finding the relevant lessons within the subject we located; external accessibility eases the process of finding the relevant subjects in the knowledgebase. Both access points not only save us time and make our work more efficient, but also greatly affect our ability to approach these lessons in the first place, ultimately making our work more effective.

Thus, we have learned that we should aim to manage each lesson as an independent entity. We also learned that we want to assign attributes and values to each lesson.

Before we continue, some further explanations are required. The terms *attributes* and *values* were used together several times in this chapter, and though it may seem as if they are synonymous, they are two distinctly different terms. Attributes are defined as characteristics. In a software project's knowledgebase, we may define attributes such as sector, project scale, risk level, technology and software development environment, or programming language. These are attributing *families*, each including a possible set of values. These values are possible elements of sets of attributes. The attribute *sector* probably will include values such as finance, public, and services; the attribute *project scales* may include values such as small, medium, large, and mega-project. Different attributes may have similar names for their values, yet their specific meaning is derived from the context in which they are used—that is, the attribute they are describing. The *risk-level* attribute, for example, will include values such as low, medium, and high; yet the medium value in project scale and risk level refer to two totally different characteristics of the same project.

With these terms defined and differentiated, let us proceed with describing the structure of the knowledgebase. Each knowledgebase, as previously defined, is a database containing knowledge. It is based on text records, where each record is composed of two groups of fields (record columns).

The first group of columns consists of *universal* attributes and values. These attributes are common to every lesson knowledgebase in an organization, usually identical as well to those of other organizations. They are somewhat technical and assigning these attributes does not require much thought. Two typical universal attributes are date (when the lesson was learned) and contributor (name of employee or team suggesting the lesson).

The second group of attributes varies from knowledgebase to knowledgebase. These attributes, known as *content-based attributes*, define the world of content in which the engineering, marketing and sales, and operations are conducted. Not only do the lessons themselves differ from one another, but also the attributes, and accordingly the values of these attributes, are different because they were derived from different disciplines (e.g., engineering or sales).

This differentiation is based on Bob Boiko's ideas presented in the *Content Management Bible* (2002), which refer to four types of metadata. To keep things simple, I refer to only the two main types: universal attributes and context-based attributes.

We have established an important rule (and it might go unnoticed if not stated explicitly): the values of a debriefing session and the lessons learned within this session are not identical. A debriefing session's values, and those of other processes from which we have learned (described in Chapters 2 through 5), define the contexts in which the lessons were learned. The values of lessons, however, define the context in which each lesson is applicable (i.e., should be used in the future). If a lesson was learned following an orthopedic surgery, it also may be applicable to heart surgeries.

We must assimilate the understanding that these are two different sets. We must understand that the values of each lesson learned are critical because they assist us when searching for lessons with the hopes of implementing that knowledge in a new situation. These values define the context in which this knowledge can and should be used to improve future life or business decisions.

The Structure of the Knowledgebase

A lessons-learned knowledgebase is a combination of four components: (1) the lesson's body, (2) context-based attributes, (3) fixed attributes, and (4) hyperlinks and attachments.

The lesson's body: A short title and the lesson itself (Table 7.1).

Context-based attributes: As you might remember, context-based attributes were introduced earlier accompanied by an example in which the main context-based attributes were sector, risk level, and project size. Unlike the first component of the knowledgebase (the lesson's body) for which we defined two integral components (title and lesson), this component does not have a fixed list of attributes. We do not even have a fixed number of columns. One knowledgebase can be defined by two context-based attributes (e.g., region and product) whereas another knowledgebase can be defined by three context-based attributes (e.g., material, process, and machine). As a rule of thumb, two context-based attributes are the minimum, and five is the maximum (Table 7.2).

Two-stage attributes can be defined, as appropriate (e.g., process and subprocess; region and state). Each organization should define its own set of context-based attributes. Furthermore, each organization can hold several knowledgebases, each serving a specific discipline. For example, an organization can hold one knowledgebase for marketing and sales; a second knowledgebase for engineering,

Table 7.1 Example: The Lesson's Body

TITLE	LESSON
Special Holidays	When defining a timetable for a project held in regions with a large Muslim or Jewish population, the Muslim and Jewish calendars should be reviewed. Any holiday found should be taken into consideration. The Muslim and Jewish calendars are not fully synchronized with the Gregorian calendar, and the dates of these holidays vary from year to year because there are less than 365 days in these calendars.

Table 7.2 Example: Context-Based Attributes

ACID	PROCESS	TITLE	LESSON
BPA	Precipitation	Adding antisolvent	For precipitation of bisphosphonic acids (BPAs), it is recommended to add the antisolvent at elevated temperatures (reflux) to reduce the level of residual silicon oil in the product.

safety, and manufacturing; and a third knowledgebase for IT (software development and maintenance). Naturally, each knowledgebase will have different context-based attributes. Mixing all these lessons in one knowledgebase and setting the same context-based attributes simply does not make sense. Think of your own libraries. Would you put company offers in the same library as the manufacturing bills of materials? Although both are documents, you probably would not sort them together because they are required for different processes and are thought about in different terms. We should follow the same logic when handling lessons. Although all knowledge in this case shares a common format (i.e., "lessons"), this is not reason enough to mix all the lessons together in one knowledgebase. It will be more effective to store each group, along with the context-based attributes that define it, in a unique knowledgebase.

Fixed attributes: In addition to the body and the context-based attributes, a lessons knowledgebase should include fixed metadata. The fixed metadata, as may be understood from its name, is identical across the organization and will include some or all of the following: date of creation, date of update, next validation date, creator, and sensitivity.

Some of these fixed attributes are self-explanatory. We all understand why a creation date is required; the same is true regarding the date of an update and the name of the creator. But two additional fixed attributes are important and should be clarified: next validation date and sensitivity.

Next validation date is an attribute that assists the knowledgebase manager in deciding when the lesson should be revalidated. Chapter 11 describes the role of such a manager who would be in charge of the quality of these lessons and other best practices for storing them. Why is such an attribute required and even considered important? When learning a new lesson, a good practice, or any other learning piece, we cannot guarantee they will be valuable forever. Times change and what was precious yesterday may be useless tomorrow. This is where the *next validation date* attribute plays a role. Do not be mistaken; newly learned knowledge that we predict soon will be irrelevant, should not be addressed; it should have been filtered out of the knowledgebase in the first place. The knowledgebase is to exclusively contain knowledge that will be valuable long enough for the organization to derive some benefit. To make the lesson-learning process cost effective, the profit the organization makes by utilizing this lesson must be much higher than the cost of adding it to the knowledgebase and handling it. When adding a new piece of knowledge to our knowledgebase, we must ask several questions. Will this knowledge be relevant in the long run? And if so, when should we reassess its relevance? Does its relevance depend on the available technologies and therefore should it be revalidated every 2 years? Is it a managerial lesson now that it is seemingly infinitely true, yet can possibly become outdated 5 years from now? The *next validation date* attribute ensures that we keep the knowledgebase accurate and updated over time.

Rest assured, the item will not vanish automatically after said date, and it will not automatically be archived. We do not regard this date as an expiration date. Its purpose is to help the manager test the knowledge and revalidate its value from time to time. The period assigned can be a year from last update when dealing with rapidly changing disciplines, or much longer when dealing with slowly changing ones. As part of his or her position, the knowledgebase manager can search the knowledgebase for items nearing or past their validation dates, and decide whether to update their content, pass them to an archive, or leave them as they are. Although we do not consume medicine or food after its expiration date has passed, one must differentiate between an expiration date and a validation date. A validations date's enforcement is more flexible. It is merely a reminder to assess a recommendation.

A tip regarding this attribute: usually, the period between last update and validation date can be set for a specific knowledgebase. If we decide to revalidate lessons and practices in our marketing and sales knowledgebase after 2 years, this probably will be the case for most items in the knowledgebase. It therefore could be set as the default date—every item would be tested 2 years from the date it was added or last updated.

Every rule has exceptions, however. The rule might be waiting 2 years before rechecking 90% of each item's content. Yet there are some specific lessons and practices for which we will want a longer or shorter period. That is why it is important to adjust this attribute for every record placed in the knowledgebase and not calculate its value based on other fields. Handling and adjusting validation dates enables us not only to build a quality-based knowledgebase but also to maintain and update its content, keeping it fresh and relevant.

In addition to *next validation date*, there is another fixed attribute: *sensitivity*. Sensitivity, like a validation date, is an attribute one could manage without. Yet utilizing this attribute adds value to the knowledgebase and specifically makes its usage much more effective.

To understand what this attribute symbolizes, let us step back and view our users. Let us imagine, for example, that we have a marketing and sales knowledgebase. The users are marketing and salespeople. Let us also assume that among these users are Marie and Joe, two salespeople. Marie (age 27) is a young saleswoman who has only started her first sales job 6 months ago. Joe (age 40), on the other hand, has been working in sales for the past 15 years. When we come to design our knowledgebase, who do we have in mind? Is it Marie? Perhaps it is Joe? Maybe a third party? If we design a knowledgebase suitable for newcomers, it will include many recommendations and practices that may seem trivial and useless for veterans. After reading some trivial instruction, veterans probably will stop using the knowledgebase. They will find it more of a nuisance than a work tool. If, however, we decide to design the knowledgebase with Joe in mind, we probably will include only the most innovative practices and unique lessons. Again, we would miss a large portion of our target audience because we want to insert a large chunk of know-how and how-to information into the knowledgebase for Marie. Yet we would not do that if we had only Joe in mind.

This is where the sensitivity attribute plays a role. This attribute can be measured, for example, on a scale of 1–5. That means that its values are 1, 2, 3, 4, and 5. Assigning a low sensitivity level to a lesson means that more users can view it. If we assign 1, everybody can view it. If we assign 5 to a lesson, only those who define themselves as beginners in this discipline will view the items when going through the knowledgebase or searching within it. Levels 2, 3, and 4 would enable the moderate levels of important, average, and somewhat useful accordingly. Assigning a sensitivity value to a new lesson is not as complicated as it might seem: simply set the level 3 as your default option, decide upon adding any new lesson as per its innovation or importance, and adjust the attribute to the appropriate level.

We have described the first three elements of the knowledgebase: title and the body of the lesson, context-based attributes, and fixed attributes. The last component to review is the hyperlinks and attachments.

Hyperlinks and attachments: The documents containing the summary of the debriefing session, mentioned earlier in this chapter, were not forgotten. They serve as second-level content for users who wish to examine and understand the roots of a lesson. These summaries are added to the knowledgebase as attachments. The attachments or links can fill additional roles, such as the following:

- Attachments or hyperlinks can direct the user to pictures or diagrams, explaining the lesson or practice.
- Hyperlinks can connect to operational management systems, shortening the transfer between the lesson and the computing environment in which said action is to be implicated.
- General or specific attachments or hyperlinks can explain why the user should apply the lesson or practice, demonstrate how to use the information, or make the derived action itself easier to perform.

That is basically it: a knowledgebase is a set of lessons, practices, and other types of knowledge presented as conclusions, with each including a title, values assigned to context-based attributes, values assigned to fixed attributes, and attachments and hyperlinks.

The Lessons

We have thus far described nearly every component of the knowledgebase, yet the most important component of all, the lesson itself, was somewhat neglected. The lesson itself requires a description. Lessons are not as easy as they seem.

Analyzing the situation and considering recommendations requires brains, or at least a well-defined methodology. We already elaborated on those. The debriefing methods described in Chapter 4, and the quality processes described in Chapter 5, are designed precisely for this purpose: assisting us in producing the best lesson possible.

Yet that is not enough. I have participated in and observed the results of hundreds of debriefing sessions in my life. Most lessons did not seem revolutionary; not even close. They usually look like trivial declarations, overused slogans, and clichés. Do not misunderstand me. The debriefings I have read, in most cases, were written by intelligent people. The participants at most debriefing sessions included stakeholders. So what is the problem with our lessons?

Ecclesiastes 1:9 states: "There is nothing new under the sun." Is that indeed the case? Is there nothing to learn? I beg to differ. Even if all rules are now known, this is hardly relevant to our learning process. The knowledge we seek when debriefing or attempting to produce lessons from any other source is derived from our interpretation of existing rules implicated in a specific organizational context. This interpretation is what is missing with most lessons. They are written as managerial or complex expressions. When reading these cryptic lessons, we just might agree with Ecclesiastes. True, it is important to keep lessons and recommendations general and not confine our findings to specific situations; rather it is better to address all situations to which the lesson at hand may be relevant. The saying (or writing), however, has to be concrete and specific; it has to provide added value. General lessons should not be abstract.

Here is a lesson I bumped into while working with a large high-tech, defense industry-based organization: "When operating complex projects, it is critical to allocate a high professional technical and managerial team."

Reading the lesson, we might wonder what it is trying to teach; nothing here is new to any of us. But, as it might have occurred to you, this specific organization may have conducted a highly complex project with an inexperienced team. The organization debriefed on said project, and the outcome was this obvious, general conclusion. To be honest, this probably happens to all of us. We assign some tasks to inexperienced people. This is not because we are unaware of the importance of a good team. It is just that sometimes, we have too many assignments on our hands and every project requests "the most professional team," which is an eternally scarce resource. A better lesson in this case would be as follows: "It is always preferable to allocate the project's technical and managerial teams with experienced people for required tasks. When such is not an option, peer assistance and close monitoring should be defined, assuring the project is executed with defined resources." Such an elaboration is referred to as "purification"—that is, processing the lesson's content and syntax, ensuring its value, ensuring that the lesson is concrete yet relevant for several possible situations, and actionable, rather than abstract.

Each lesson added to our knowledgebase must be purified. It should be presented as a sentence or paragraph that is easily understood, was agreed upon, adds value to the user, and is concrete. It cannot contradict previous lessons or repeat what already has been learned and written. Repeated lessons should be merged; in cases of contradictions, both the new and existing lessons should be reviewed and the contradiction should be resolved.

Following these instructions may require time and effort, yet it is worthwhile and ultimately cost effective. The time spent on this process is what upgrades the lessons and makes the knowledgebase an important organizational asset.

Congratulations! We now have a knowledgebase. We have added a second floor to our world of lessons; now we not only produce new knowledge, but also manage it in a purified knowledgebase:

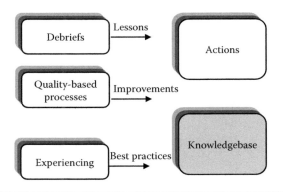

The lessons and practices cycle—the knowledgebase.

So what is next?

We have two more stages in our life cycle of lessons and practices, in addition to creating the knowledge (Chapters 3 through 6) and managing the knowledgebase (Chapter 8). These final stages are proactive: transferring the knowledge, making it accessible to the user who might need it. This will be dealt with in Chapters 9 and 10.

Reference

Boiko, B. *Content Management Bible.* New York: Wiley Publishing, 2002.

8

Embedding the Lessons in the Organizational Environment

David is driving down the road. In a few minutes, he will enter the offices of Radan Engineering. He has made an appointment for a negotiation meeting with the potential customer and is keen to earn this business; David wants to close the deal. He does not yet know how to bridge the gap between what his company has proposed and what he presumes the customer is willing to pay. The company surely can benefit from this sale, and so can he. But can he pull it together? How can he compromise without the customer sensing that he really needs this deal and taking advantage of the situation?

The phone beeps. He parks his car. He still has 15 min to spare before the meeting. He checks his phone. Sarah was looking for him. He calls back, and still having a few minutes to spare after answering her question, shares his concerns with her. "You know," Sarah reminds him, "costumers releasing products for over 2 years are offered up to 15% discount if the customer is willing to participate in the beta program for the next release of the product. This discount may be offered even when no program of the sort is scheduled for the time being."

David recalls that about a year ago, Sarah told him about a brainstorming session in which she participated as part of the product management department. They had a problem convincing companies to move from existing versions of products to new ones; it was especially challenging to convince the large organizations, from which they would benefit most. The company had a strict policy against discounts because it was concerned that discounts might seem disrespectful to other clients. Also, the company reasoned, once you give one discount, there is no end to it. In this specific situation, however, management felt comfortable with the idea of discounting. It was backed by solid reasoning: the company would benefit from this deal and this specific discount probably would not harm the company's reputation. This idea was implicated in other deals as well and was effective.

Was this knowledge documented? Of course it was. Was it shared? It is safe to assume it was, but it was forgotten. Is it reasonable to assume that before every sale, David and all the other salespeople would review ideas developed and reported in the past, searching for elements relevant for their deal? Probably not.

Storing the organizational lessons and good practices in a knowledgebase is an improvement over not recording any of this information, but it is insufficient for most organizations. Relying on coincidental chats, like the one just described, is not a reasonable option.

Any organization that invests in debriefing and constructing a knowledgebase should go one step further. The organization should design means to embed its knowledge into the organizational environment, thus creating a work environment in which knowledge is near and accessible to the relevant workers when needed. As Frank Leistner, who served as the knowledge officer at SAS Institute has stated, "The best way to keep knowledge from leaving the organization is to embed it" (2010, p. 90).

Some readers may believe that only magic can conjure such a solution. Granted, some magicians may have some insights and advice about this approach, but that is not this book's main focus.

This book offers conventional, easily followed steps that push the knowledge closer to the employee. The next section reviews typical examples.

Templates and Forms

Templates and forms are wonderful tools to embed knowledge. In many cases, people filling in forms and templates deal with the creation process of new knowledge, so the chance is good that they may be in need of lessons and good practices concerning the development of this new knowledge. Many of us use templates when writing proposals, preparing project work plans, and writing technical specifications. We also may use templates when designing tests, requesting the organization to approve a service provider, as well as in many other core organizational processes. Organizations invest in preparing forms and templates for these processes, leveraging the ability to reuse knowledge. They make this investment mainly by setting the specific steps and topics that should be included in each process. Embedding lessons and good practices into these forms and templates elevates organizations to the next level of knowledge reuse. We not only define what should be part of the process but also explicate how to best perform these tasks. Obviously, the consequence of some lessons is to change the form or template (e.g., adding a subprocess). The following solutions instead refer to newly added knowledge. This is knowledge that employees should be exposed to when using the template or completing the form, while allowing them the opportunity to choose which elements of this knowledge may serve them in the specific situation.

Suppose a project manager is preparing a new project's kick-off. The project manager uses an organizational template that assists the manager in choosing which aspects should be presented as part of the kick-off meeting. The kick-off template may include topics such as stakeholders, teams, work plans, and main risks. It is only natural that the knowledgebase will include lessons regarding these topics. More specifically, it should include lessons relevant to this preliminary stage of the project.

The idea is not to merely add more sentences into the templates and forms, further describing the knowledge. Such a solution may cause information overflow in the templates and forms. Furthermore, this approach is not dynamic and requires the addition of a new sentence for every new lesson that addresses managerial issues during the kick-off stage. Such additions may produce exhausting templates. A much simpler solution is to add a single word or icon at the end of each line, including a title of the template or form directing the user to relevant lessons.

The symbols (Figure 8.1, on the right-hand side under each title) are hyperlinks to the knowledgebase. They do not link to the full knowledgebase of lessons; rather they point to a focused set of lessons relevant to the specific title (technically, this solution can be implemented easily by performing queries on the knowledgebase). Users completing the template will know that within one click, they have access to all of the organizational lessons relevant to the issue with which they currently are dealing. This technique leaves the original template or form clean, not only enabling the employee to concentrate on what has to be written

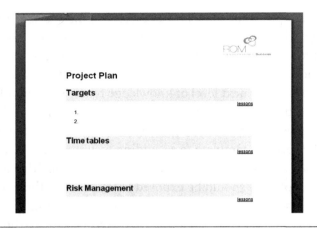

Figure 8.1 Project template.

to complete the relevant process, but also allowing the organization to expose the user, each time the hyperlink is activated, to all updated lessons. This includes lessons added after updating the form and adding the link. The nature of such hyperlinks is that the query is processed in real time when activated by the user. Therefore, all updated lessons learned and good practices are included in the results list.

This is the first step in embedding knowledge in the user's environment. The knowledge is not force-fed to employees; rather, it is brought closer. The knowledge will be accessed and used more easily. It is assumed that most employees want to succeed in their work and would appreciate the opportunity to review lessons learned and good practices. This is true as long as the cost of remembering to search for the lesson, accessing the knowledge, and understanding and applying it is not too high.

The following are additional ideas about how knowledge can be embedded in the organizational environment.

Training

Training is the main learning opportunity offered by most organizations. But we do not learn only at work. We learn all our life: at home, at work, when hiking or camping, or when pursuing any other activity. We learn when performing any activity in which we take part. Yet training is a somewhat different form of learning, as it includes proactive learning. We stop our daily activities, sit in a classroom, and acquire new knowledge (or at least intend to do so). It is only natural to include organizational lessons and good practices as part of any training course dealing with these topics. Training usually is composed of several layers; some basic, others more advanced. Each layer (especially the latter) includes lessons as part of teaching the good "know-how-to-dos." One stage in which we should impart this knowledge is the initial development of any new training project. Those in charge of developing courses usually seek all of the relevant knowledge and obviously appreciate knowledge extracted from the lessons knowledgebase if they happen to stumble upon it. Rather than rely on coincidence, the training department manager should teach and instruct the person in charge of the training development to explore the lessons knowledgebase and choose appropriate lessons learned and good practices to be taught as part of every course. Even this recommendation, however, is far from sufficient. The problem with this suggested technique is that training courses usually are developed, redeveloped, or fully updated only once every few years. In the meantime, many new lessons could have been learned; many good practices should have been acquired. Unfortunately, these additions will not be reflected in the training, even if the lessons initially were embedded in the course. To address this problem, two additional techniques can be implemented to help us reuse the knowledge we already have developed. Note: these techniques are complementary to the first technique (searching the lessons knowledgebase and sharing lessons learned) and are not suggested as a replacement.

The second technique is easy to implement. Simply add a session discussing interesting lessons to each course. This can be addressed by adding one slide to the presentation. The presentation slide could include a single line: a hyperlink to the lesson's knowledgebase. The hyperlink can be specific, querying lessons related to the topics of the course. Every time the hyperlink is activated, the students will be exposed to all relevant lessons learned, including the new lessons added after the course was developed and surely including the updated syntax of each lesson (assuming some were updated since they first were learned). The teacher can either go through the whole list of lessons or choose (either independently or following a classroom discussion) which lessons should be discussed this time.

The "interesting lessons session" technique has two advantages. First, the lessons taught are always the most relevant ones; no detail is missed. The second advantage is derived from the way the lessons are presented. In the traditional method, students were aware only of the contents of the lessons, and sometimes they were not even aware that it included lessons learned. When using the "interesting lessons session" technique, the teacher then exposes students both to the lessons and to the idea of a knowledgebase. In so doing, students are taught that they can and should access the knowledgebase in the future for ongoing needs.

The third technique utilized to demonstrate how to embed knowledge in training refers to exercises. These exercises usually are an important component of most courses. The students may be struggling with a challenging problem regarding what they have learned (as well as what they have not learned). This might send them in search of relevant ideas about how to solve their problem based on various sources, including lessons learned and good practices stored in the knowledgebase. This technique prepares them for their job and encourages them to proactively approach the knowledgebase on their own will. I recall a time this technique was used on a knowledgebase that was not fully launched, which meant the number of authorized users was still quite limited. The day after the training was completed the lessons knowledgebase manager was flooded with requests to join and receive access to the new system. Those who graduated from the training course returned to the base (this was a defense intelligence unit) and shared the idea of the knowledgebase with their colleagues, who in response contacted the manager to hear about this new "well" of knowledge and request access to it.

The last two techniques are relevant mainly for organizations and working environments in which templates and forms are used and training courses are held. Before we examine organizational methods, we will discuss a simple method that is relevant not only for professional organizations but also for home. In fact, I use it on a regular basis.

Making It Hard to Make the Mistake

If making a mistake is difficult and uncomfortable, the chances we will make it are lowered. This statement might seem unrealistic, but memorize the following explanation and internalize it because it actually works.

I will begin with a personal example. Some years ago, my husband took piano lessons. His teacher, who was a brilliant musician, was an unorganized fellow. He would come to our home once a week, around the end of the day, and rarely remembered to leave our house with everything he came with. One evening, he arrived with a milk carton in his hands. He told us he bought it on the way because he ran out of milk at home and the stores would be closed by the time he finished giving the lesson. He asked for our permission to put his milk in our refrigerator. My husband, who knew his teacher well, realized that the chances this milk would reach the teacher's home that night were nil.

"Give me your car keys," he offered to the teacher.

"Why?" The teacher inquired, puzzled.

"I will put them in the fridge as well, beside your milk," my husband explained. "That way you won't forget to take the milk. The furthest point you will reach without your milk is the door of your car."

Since that night, my keys have been chilled often, and it always has been worthwhile. Car keys, as well as cellular phones, are objects we find hard to be without for an extended period. Therefore, putting other important objects near them can help any one of us who sometimes forget.

Daniel Levitin, in his book *The Organized Mind* (2014), gives a similar example: "If you're afraid you'll forget to buy milk on the way home, put an empty milk carton on the seat next to you in the car" (p. 85).

As we learned earlier in this book, generalization is an important feature of learning lessons. The following story exemplifies this point. One day, after I finished meeting with a company, I drove one hour home only to realize that I had left my ID card, which I was required to hand in when entering the company, back at the company's reception desk. I then decided to implement the car keys method. Since then, whenever I am visiting a company and am requested to hand over my ID card and receive a guest ID, I always attach the company's guest card to my car keys. Just last week, I found myself confidently walking to my car, taking out my keys, and noticing I was carrying some "extra luggage." It took one glance for me to change direction and return my guest ID (and, more important, retrieve my own ID).

This "car keys in the fridge" method can be implemented in many other formats. In the past few years, on numerous occasions spanning almost every country, people have forgotten their infants in their cars. We tend to continue with the routine, even when change is required. People just keep on driving, oblivious to the fact they forgot to drop their child off. During the sweltering summer days, this mistake can yield tragic results. In 2013, a shocking 44 children in the United States alone were reported dead from suffocation resulting from this horrific phenomenon. A campaign intended to minimize this recurring tragic error suggested various ways to prevent forgetting children. One method suggested putting the driver's bag or wallet near the infant, instead of on the passenger-side seat. This idea serves the same purpose as putting our keys in the fridge. It might not prevent us from driving directly to work, but in most cases, it will ensure that when we reach for our bags before exiting the car, we will see that no baby has been left behind.

Another safety-related problem deals with the way some cutting machines are designed to operate. After too many occasions of workers accidentally sticking their fingers into cutting machines, some machines have been redesigned, again to help us prevent this dangerous mistake. Many cutting machines require the person operating the machine to press different buttons simultaneously. These buttons were placed far away from each other, so that one can simultaneously press them only by using both hands. This is another example of making it really difficult for average workers to make a mistake and cut their fingers. Years of various attempts to educate people not to stick their fingers into the machine yielded unsatisfactory results. Engineers had to make it harder and more uncomfortable to actually make the mistake to provide a safer work environment.

As emphasized, this technique has many forms and is quite useful. But like nearly everything in life, applying this technique is not as simple as it would seem. It turns out that finding what can be done to make it really difficult for us to make those mistakes is the truly difficult task. This is why we need even more techniques.

Checklists

We all use checklists. Even when we go to the supermarket with a list of groceries, we are using a checklist; we stroll down the aisles adding goods to our cart and checking off the relevant item on our list.

Checklists are popular in organizations as well. They are used to ensure that we follow the process step by step, in the order found best by the organization. These checklists serve both as reminders and as supervision and control records. In the context of organizational lessons, checklists have an important role.

In some cases, as a result of a debriefing and learning session, a new checklist is created. Even in my home, I keep a small checklist in a discrete corner of one of my closets. This small checklist, which I use at least once a week, already has saved me from forgetting things I should do. This might not come as a surprise: I made this list after making the same mistake not once, but twice. These mistakes were a result of forgetting.

Creating a new checklist may seem like an appropriate solution for almost every lesson, yet it is not. Too many checklists give the impression of bureaucracy and technocracy. Furthermore, most processes are not appropriately defined. In the rare occasion a process is well defined, the checklist still may be perceived as a burden rather than an aid. When checklists fit the situation (and we are not overwhelmed by them), however, they do their job wonderfully; in some cases, they work like magic, fully embedding the lessons learned into organizational behaviors.

We are even luckier in cases in which a checklist already exists and the new knowledge simply can be inserted. In these cases, a checklist is indeed the right solution. No change management is required, as no new process needs to be handled.

If the lessons fit a current checklist, or if the lesson has important implications and requires a new checklist, we can feel blessed. The chance that these lessons actually will be used is quite good. If this is not the case, however, additional solutions must be considered. We have discussed some ideas already, including templates and forms, training, and making the mistake harder. Let us consider some additional ideas.

Presenting Lessons in Search Processes

Disclaimer: Before we discuss this method, note that it does not suit every situation. It is an efficient method when applied in an organizational context, when the team managing the lessons learned can

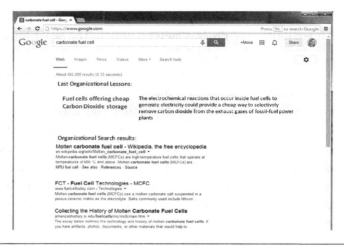

Figure 8.2 Lessons embedded in search results.

initiate changes in organizational computing, and, specifically, in the implementation of the organizational search.

In such cases, the lessons can be embedded as "Google ads." This is a small area within the window that appears above the search results. This area would feature lessons that are associated with the subject of the search. Figure 8.2 illustrates such knowledge embedding; the lessons are added above the other results and are slightly highlighted.

The lessons are separated from the original list of search results. This is not because they cannot be part of the search results themselves. They are separated for the same reason that ads are highlighted; we want them to be noticeable. Employees search for what they need the most: information and knowledge. That is exactly the stage at which the relevant lessons learned should be offered. And that is exactly what this method provides: the worker searching for knowledge receives it along with the most important organizational knowledge: the lessons learned and organizational best practices.

Online Help

Another technique somewhat similar to embedding lessons within search results is embedding lessons in operational computing systems. This technique may be performed in many ways, yet one of the most popular methods suggests embedding the access to lessons as part of the online help. For many years, we did not give the online help field the attention it deserved. Online help was something that software producers had to provide as part of a system. It was a minimal feature and, in most cases, rather operational, providing explanations for trivial content on the screen. It explained how to fill in the fields presented on the screen from which the software was originated and showed which buttons should be chosen in each case. Lately, systems have become more user-friendly and more intuitive to operate. In some systems, online help has been abandoned. There is almost no need for operational guides. In organizations in which online help still exists, however, it can serve as a useful placeholder for lessons learned and good practices. Instead of thinking of the help function in terms of a set of explanations about how to operate some action, it can be leveraged. It can include, besides the basic explanation, various types of useful knowledge. It may contain frequently asked questions, presented with their answers, helping users find their way when they are unsure how to execute a specific action; it may contain contact information as to who to address if any unresolved questions remain; and it also may include relevant learned lessons that somehow are related to the content of the displayed screen.

The lessons could be embedded within the page itself, as part of the help function's explanation. This, however, is less convenient as it implies that every time we define a new lesson, we need to copy it to the relevant help pages. To avoid this inconvenience, instead hold a dynamic list of current relevant lessons. Each time the list is displayed, it reveals the relevant lessons stored in the knowledgebase. The list presents matching lessons and online help screens based on the metadata. This way, we not only help our employees perform the desired action but also enable them to perform it better, taking organizational wisdom into consideration.

Processes

Embedding lessons within the processes in which we need them is the classical method of embedding knowledge. For example, we might have defined lessons about the circumstances in which it is not recommended to work with subcontractors as well as lessons about the advantages of working with small contractors. The best solution in this type of case is to embed these lessons in the organizational process of choosing and approving a new contractor or purchase order. A defined lesson such as this can affect a work process in one of two ways. If the lesson better defines the current process, the change will be embedded into the process itself; in such cases, we do not need to add a new item to the knowledgebase, and we are not required to embed this knowledge. We can assume that the people who were not part of the lesson creation will not be aware that this definition of the process was derived from a lesson-learning process. That is okay. In this case, all that matters is that the process is improved; not that some lesson was learned.

In other cases, however, the lesson does not change the process itself, but rather adds some information guiding how it can best be processed. In these cases, the lesson should remain in the knowledgebase. If some form or template is related to the process, it can include a link to the relevant lessons, stored in the knowledgebase as defined at the beginning of this chapter.

Work Procedures and Guidelines

Last but not least, the work procedures and guidelines of the organization can be used to embed knowledge. In many organizations, they still serve as the only "home" for lessons learned. The classic approach to debriefing ends with writing a working procedure or guideline. If a relevant working process or guideline already exists, then it is updated with the new knowledge. It is not a coincidence that this method is presented as the last method, and not the opening one. Work procedures and guidelines suffer from several problems and definitely should not be the first solution to attempt knowledge embedding. Procedures and guidelines suffer from unpopularity. On too many occasions, employees do not use these as sources of important organizational knowledge, even though this is exactly what they are supposed to be. The reasons they are not used vary, and this book may not be the place to raise the issue. Let us just agree that it is unwise to solely rely on an unpopular and rarely used channel for this knowledge, important as it can be.

Frankly, the unpopularity of procedures and guidelines is not the only reason to discourage their use as the only or even main channel through which lessons learned are delivered to those seeking this knowledge. Procedures and guidelines represent a formal source of knowledge that instructs workers about what must or must not be performed. As we learned in Chapters 3–5, lessons are much richer. Lessons may include recommendations, things to consider, and other types of informal knowledge. These are, admittedly, not strictly defined as knowledge and, as such, have no place in the world of organizational procedures and guidelines. So, in cases in which the lesson is a formal one, and a procedure or relevant guideline exists, it should be embedded with this information. When a relevant procedure or guideline exists, we will avoid including the lesson in it to prevent confusing the mandatory with the optional. What we want to do instead is point to the lessons—for example, add a paragraph to the procedure or guideline, providing the user with a link to the "organizational-related wisdom" stored in the lessons knowledgebase. Separating the lessons from the document helps users differentiate between the mandatory and other important knowledge (i.e., lessons). In other cases, when the lessons are not related to the work procedures and guidelines, we may use one of the other methods presented earlier. We have not presented a full list of options in this chapter; rather, these are representative. Everyone implementing lessons, at home or in an organization, can develop many more ideas based on the structures described in this chapter.

Summary

Embedding organizational knowledge sometimes may seem irrelevant to those who deal with lessons. Nevertheless, it is an essential stage in the life cycle of lessons and practices. Without this knowledge, we might have lessons learned, but not lessons applied.

Thus far, we have discussed debriefing (Chapters 3 and 4), described other quality-based processes (Chapter 5), and specifically elaborated on experience (Chapter 6) as additional sources of knowledge. We discussed actions and defined a knowledgebase storing all this information (Chapter 7) and added a third stage of embedding knowledge in the organizational environment (this chapter). These are represented in Figure 8.3.

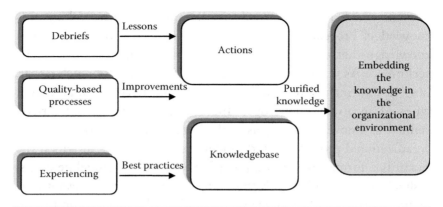

Figure 8.3 The lessons and practices cycle—embedding the knowledge in the organizational environment.

We need to address one more stage to complete this life cycle: requesting the knowledge *before* action. This stage is described in Chapter 9.

References

Leistner, F. *Mastering Organizational Knowledge Flow: How to Make Knowledge Sharing Work (Vol. 26)*. Hoboken, NJ: John Wiley & Sons, 2010.

Levitin, D.J. *The Organized Mind: Thinking Straight in the Age of Information Overload*. New York: Penguin, 2014.

PART IV
RETURNING TO THE PREVIOUS DAY

9
Requesting the Knowledge before Action

Greek mythology tells us the story of Cassandra, the daughter of King Priam and Queen Hecuba of Troy. Cassandra was a beautiful princess and Apollo, who wished to seduce her, granted her the power of prophecy in exchange for her promise to love him. After she turned him down, he decided to punish her. Apollo cursed Cassandra's prophecies; although her prophecies still would be accurate, no one would believe her.

One may not understand why this curse is viewed as such a harsh punishment. Yet Greek mythology offers an important insight: knowing the future and attempting to prevent disasters to no avail is a curse. Cassandra went insane.

This also is one of the most frustrating things that can happen to us when managing lessons: we perform tasks, and the results differ from the desired results. So we bother to take the time and debrief. I use the word "bother" intentionally as debriefing is not easy. It is time consuming, we are all in a rush, and most of us already are working on other tasks. It is thought of as exerting, as effective lessons usually do not just pop into one's brain. And above all, it requires us to be sensitive, as we must invest energy to prevent some people from feeling criticized while also preventing other people from finger-pointing.

Following this tiresome ordeal, we begin working on the lessons. We generalize them and review additional cases and contexts in which the lessons are applicable. We validate the lessons to ensure that new lessons do not contradict existing ones. We pay attention to wording, verifying that the phrasing is professional, politically correct, and organizationally wise. We think of ways to embed the knowledge within the environment and its processes. We then feel optimistic, assured that now things will be better.

But, once all this work has been completed, if the knowledge is not being applied and new groups in new projects and new circumstances are repeating the same mistakes or making similar ones, then there is no other way to put it: it is frustrating.

It is one thing to err once; it is a totally different story to repeat (or watch others repeat) similar mistakes multiple times. Organizations have a hard enough time operating efficiently without having to reinvent the wheel on a daily basis. (Although, I will admit, this has happened in our own organization more than once and even more than twice!)

What can organizations do to prevent such a phenomenon from occurring, or at the least, what can they do to ensure that it occurs infrequently? Chapter 8 discussed embedding new knowledge in organizational habits and routines. Doing so is fine, and every organization should aim to maximize such processes. Yet maximizing embedding processes is far from sufficient. We do not have the means to embed this knowledge in all situations of life in which they might come in handy. Furthermore, organizations are in constant motion. So, how do we bring new knowledge into the minds and actions of people who already are preoccupied with other matters? How do we ensure that the new knowledge will serve them here and now, and not only serve future needs? How do we ensure that they will be notified with relevant knowledge before beginning new processes?

How do we cause them to request knowledge *before* acting?

No single solution is correct for dealing with this challenge; there is no absolute recipe that one can blindly follow. We cannot predict all possible situations in which the new knowledge might be required, and therefore we cannot suggest it to all of the employees who might need it.

Several techniques have been proven useful, however, if adapted to the organization's culture and to the context of its corresponding processes and activities.

Nick Milton, in his book *The Lessons Learned Handbook* (2010), dedicates a chapter to this issue, suggesting three main possible solutions. The first solution involves broadcasting new lessons and improving processes; thus, by publishing a blog or newsletter, updating people via an existing channel of some community of practice or other forum, sending e-mails to distribution lists, pushing via RSS feeds, or even using a proprietary software. The second solution is to embed the new knowledge into training sessions. The third solution is to integrate a process review as part of the routine operations. These techniques are indeed helpful, yet before discussing these specific solutions, it may be a good idea to define the principles needed for an apt solution. From these principles, we can derive the best possible solutions.

Chapter 8 dealt with various ways to bring organizational knowledge closer to the user, embedding it into the existing environment. As explained, this is not always applicable because not all scenarios are apparent in advance. Even if we could predict every scenario, we do not always have the means to bring in the necessary knowledge.

The complementary path taken should be to identify means to motivate users to search for the knowledge themselves.

If we agree that users want to perform their job properly and not fail at it, it is important to understand why they (and we) will not use every piece of available knowledge that can help them make the right decisions and enable them to complete these tasks both more efficiently and more effectively.

To utilize knowledge, we must first know it exists; so this knowledge is something that has to be acquired, and as the literature teaches us, this must be done more than once. Consider a typical debriefing session taking place in any organization. A debriefing process cannot include all of the organization's workers. Even if it could, involving everyone in the process does not make sense. In many cases, when the debriefing team completes the process, it presents its results to a larger group of people. Similarly, more people will not attend this meeting than do attend it, as our resources (both conference rooms and employee time) are limited. These limitations can seem irrelevant when considering how easily results can be e-mailed to employees. So many of these presentations, representing what has been shared in some live meetings, are distributed as widely as possible. This information lands in the inbox of each employee, probably along with another 50–100 other important messages. We usually do not have the time to open the file attached to the mail; in the rare case it is opened, it likely will be read with insufficient attention, which in turn leads to users hardly remembering the lessons read, or perhaps skimming through them impatiently, in some cases even missing the point of the lessons learned.

This also happens in our personal life. Someone says something; maybe a mother tells her son that some task should be performed with care in a specific way; yet said son is preoccupied with other matters that probably are more appealing in his perspective. His mother's words therefore go unheard.

So what can we do differently so that the newly acquired knowledge is actually used? How can we deliver knowledge to the people who need it exactly when they need it?

The best situation I can imagine is that every time knowledge could be of use to me, it would pop up in a cartoon bubble above my head. I would always know what to do, here and now, and life would be easier.

As appealing an idea this might be, such technology does not yet exist. So for now we will have to settle for imperfect solutions.

The path to handling this challenge is twofold. Wherever possible, we must define a designated process that we intend to integrate as an organizational process and then recommend that the user request this knowledge. If employees cannot perform the designated process (remember: cartoon bubble pop-up mechanisms are still merely fiction), they will be expected to request the knowledge. As explained, users probably want to receive this knowledge, as they want to succeed; so we have set means for getting the knowledge close and accessible (Chapters 7 and 8) to said user. We also must see to it that users actually remember to request this knowledge, and a portion of this chapter will discuss how to attain this goal (or at least improve the chances it will be reached).

On some occasions, we can define designated processes that serve this need. Let us return to Nick Milton and his recommendation to include a "process review" before actions take place. Back in the late 1980s, British Petroleum (BP) started performing such a process, which they titled "Peer Assist." The Peer Assist program was the jewel in the crown of BP's knowledge program.

The process, later copied by many other organizations, is viewed as one of the preliminary steps of any projected planning for any event. This process is based on the notion that you nearly always can learn something relevant about the task with which you are assigned from someone else who has previous experience with a similar task. The person or team dealing with the new need reviews other activities, teams, or projects and chooses one from which they can learn the most. They schedule a session with peers (recommended duration: between half a day and a day); that is, relevant members representing the similar task. During this session, they describe their planned activity and needs to the peers. The peers then share their experience and best practices with the team and, in some cases, discuss some of the challenges related to the new situation.

What happens during this session, at least in its first two steps (describing the project and listening to relevant experiences and best practices), is obvious. The project or planning team learns from past lessons and experiences. This process takes place in a face-to-face meeting and therefore can be performed in a limited number of sessions. Only those sessions that featured similar attributes to the current task should be modeled. An organization that holds a lessons knowledgebase can take this process one step further. The team may learn, as a first step, from all relevant previous projects and activities by querying the knowledgebase. After doing so, perhaps based on the relevant lesson's contributors, it may hold a peer-assist session and learn about additional lessons, experiences, and understandings. Also, the team might focus the joint discussion on the challenges currently being faced.

Chapter 8 suggested that a hyperlink should be embedded into the lessons' knowledgebase easing the route to knowledge. However, the organization can go one step further: it can add, as part of the project plan, a section in which the project manager will have to note three lessons learned from other projects or indicate occasions that are relevant to the current situation. If no lesson is relevant, that is fine as well, but this conclusion must be justified by a solid argument. We have to assume that people want to give this process their best (so long as it does not require too many resources), so this requirement seems reasonable. Noting lessons learned from other situations seems to require less effort than not noting them and explaining why.

BAR: Before Action Review

A similar idea is the before action review (BAR). The BAR process was developed to complement the after action review (AAR) process (described in Chapter 4). AAR is a well-known technique; yet its sister technique, BAR, is much less popular. BAR is a kind of pre-mortem process. We try to predict, before handling the process, action, or event, what may happen, and what would we be saying if we were sitting in an AAR session (i.e., after, and not before the process). The BAR includes four questions. The first and the second questions ask what we expect to happen and what possibly could go wrong. The last question, naturally, asks for our recommendation for said situation. Between these stages, we ask the vital question: What can we learn from previous situations? We ask the user to seek existing relevant knowledge. If the organization has a lessons knowledgebase, this process may be relatively simple. If such is not the case, then a process resembling the peer assist can take place; the project manager can query colleagues regarding relevant issues. And, of course, all other options in between can be utilized as well (reading debriefings, reading projects summaries, and questioning veterans). Without doubt, a lessons knowledgebase is the most efficient way to answer this question, but a BAR also can be performed in organizations that do not implement the life-cycle model of lessons and good practices management, as described in this book.

We do not always have the privilege of knowing about an upcoming event in which organizational knowledge (lessons, insights, and good practices) will be useful. In too many cases, we do not know whether users will need the knowledge, so we have to direct their awareness toward this knowledge.

We have endless techniques to inform users about new knowledge; the marketing discipline specializes in such. We categorize these methods into two classes.

The first class makes use of existing channels of acknowledgment. Many of these channels exist in every organization: quarterly updates of management, team and unit monthly or weekly meetings, organizational and professional newsletters, and so on. It is not uncommon and rather uncomplicated to add a lessons corner to any one of these channels, updating the group or organization about an important or interesting lesson, and enabling the user to easily pop into the full knowledgebase, learning more information or searching for additional lessons. The Jerusalem Municipality includes a sample random lesson in the homepage of its employee website portal. Every time employees enter the portal, one lesson is out there teaching them something new or reminding them of something they once learned. Even though we might not have the time to learn from lessons every time we see one, in the cases in which we do have time, these channels remind us of this important asset.

It is also possible to share knowledge in a regularly scheduled face-to-face monthly meeting; we may ask an employee to share a lesson learned. Although this might seem like a less efficient way to transmit the knowledge, it includes a storytelling effect, making it cost effective (if we indeed have important lessons to share). Each organization has its own channels, and this knowledge can be best incorporated into each organization somewhat differently, according to each organization's characteristics and culture.

The second class includes channels dedicated to this purpose of sharing lessons. The Israeli Air Force holds an annual all day lessons conference, in which all commanders sit together and learn about key debriefings that took place and the relevant lessons learned. In the Jerusalem Municipality, a bimonthly newsletter designated for lessons and good practices is distributed to all community employees sharing three new lessons and success stories related to the lessons, and prompting employees to look for relevant lessons, while also publicizing how many lessons, experiences, and good practices already exist in the knowledgebase. The municipality is not unique for using these channels; hi-tech companies and other organizations use them as well, and statistics show that after these publications are distributed, the knowledgebase's web traffic doubles!

Last, but not least, users can help themselves. Using both classic and new technologies, users can request updates about new lessons, knowledge, and best practices. They can request alerts, notifications, or any other type of "like" that pushes new knowledge regarding matters that interest them or are important to their work into their digital workspace. When users request said knowledge, chances are greater that it is focused on their needs; that they will read it carefully; and that maybe, they just might use it.

And that is the essence of this chapter and this book: getting the user to indeed use the new knowledge, thus enabling learning and business improvement.

A HOLISTIC APPROACH TO LESSONS LEARNED

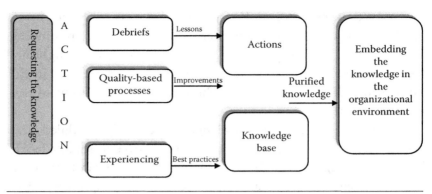

Figure 9.1 The lessons and practices cycle—requesting knowledge.

So where do we stand? Thus far, we have completed Chapters 3 through 9, the life-cycle model of lessons and good practices management, as illustrated in Figure 9.1.

We are not yet finished. For this life cycle to fully work in some organization, some additional work still has to be done.

The last part of this book deals with the task of implementing the life-cycle model of lessons and good practices management in organizations and addresses the roles that must be established in the organization.

Reference

Milton, N. *The Lessons Learned Handbook: Practical Approaches to Learning from Experience*. Oxford, UK: Chandos Publishing, 2010.

PART V
Implementing the Life-Cycle Model of Lessons and Good Practices Management

10
JUMPING INTO THE WATER

One of my favorite ways to learn a new subject is to Google it using the images filter. Of the many advantages to using this technique (you are welcome to try it), the following three are the most appealing (in my opinion).

First, images are concise, so people can learn a great deal without investing much cognitive energy (i.e., a picture is worth a thousand words). Second, images inspire; and, third, when physical space is limited (several pages or one long one), using a picture means many perspectives are offered, each contributing to a better understanding of the issue at hand.

Googling "where to start" leads to many pictures, accompanied by fine phrases, thus suggesting a variety of motivational concepts, including the following:

"The most important step of all is the first step. Start something."

"Start where you are. Use what you have. Do what you can."

"The start is what stops most people."

"A journey of a thousand miles begins with a single step."

"Decide that you want it more than you are afraid of it."

"Start by doing what's necessary; then do what's possible and suddenly you are doing the impossible."

"Start today."

These phrases (as well as many other similar ones listed in search engines) can be categorized into two groups—namely, desire and method.

The first group consists of recommendations dealing with people's (absence of) *desire* to start. Urging people to start is necessary, as people have difficulty leaving their comfort zone regardless of their destination; even if the change reveals many advantages. We fear change and usually prefer to avoid jumping into the water.

In addition to these recommendations, we find recommendations dealing with *how* to start (method). Naturally, these recommendations are general as they deal with the idea of starting anything and not with any particular beginning.

Working on the desire for change is itself an issue. It is discussed at length in Chapter 11, which is devoted to cultural change. This chapter, however, deals with the practical stage—that is, how to actually start; how to jump into the water not only remaining afloat but also having a specifically defined target to reach safely.

So be patient. Those who already have initiated many programs in organizations may find that they know much of what is written in this chapter. They may feel ready not only to jump easily into the water, but also to skip this chapter altogether and continue to Chapter 11. Please do not do so! These two chapters include many updated insights based on research, field experience, and the specific issue dealt with here: life-cycle model of lessons and good practices management.

The Decision to Start

Introducing the idea of "life-cycle model of lessons and good practices management" may not seem as simple as one would wish it to be. Management probably is flooded with many interesting ideas on how to improve business, much more than they actually can implement. Before they even start considering new ideas, they must deal with the "here and now" (i.e., all ongoing targets, challenges, and problems).

So to initiate an idea, any idea, it must fulfill three conditions in addition to being beneficial: it also must be compelling, applicable, and affordable.

There are several ways to present the considered move as compelling. Each organization has its routes, and managers should choose one suitable for their specific case. Specifically regarding pitching the idea of the life-cycle model of lessons and good practices management, the following are two typical ways found to be rather low budget yet convincing: one positive, the other negative.

The positive method is to demonstrate a proof of concept. Milton and Lambe emphasize the importance of proof of concepts as an effective method to initiate knowledge management-oriented programs in organizations (Milton and Lambe, 2016). Such proof of concept will consist of preparing a sample knowledgebase, including a handful of lessons or best practices (somewhere between 5 and 10 lessons). These lessons can be gathered via a debriefing session, captured from documented debriefing sessions, or by interviewing a subject matter expert. The lessons can be generalized, and is a good idea to also present examples of how a lesson or two can be embedded into the organizational environment. This route demonstrates how beneficial lessons can be captured effectively and efficiently, and then embedded back into organizational life. Usually, presenting the lessons reminds the managers of forgotten good lessons. Demonstrating how these could be remembered and used persuades said managers; theoretical matters become concrete and personal as they may realize just how many times they also have forgotten the lessons and even were oblivious to some.

The negative method focuses less on the solution and more on the need, demonstrating the cost of working today without systematically fully managing these lessons and good practices. One way to achieve this is to present several cases of similar mistakes that were made, even though lessons were learned and the organization was notified. Another negative approach (one might even say a cynical one) is to present similar lessons, learned time after time, that repeatedly left the organization at square one, even after investing in lessons capturing and debriefing.

Consider the following example from a real organization (actual numbers represent specific events that occurred during the years 2005–2011):

Falling from heights:

- Event number 3: Falling out a window
- Event number 25: Falling from the roof
- Event number 32: Falling out a window

Sports equipment collapsing:

- Event number 1: Basketball hoop falling off; luckily no injuries
- Event number 17: Football gate collapsing
- Event number 22: Basketball hoop falling off at a sports gym

In both cases, the examples chosen are critical to the attempt to persuade the organization to act on this information. One good example will do the job, but to reach that winning wild card, you must start with a few examples and hope they include the "one" that resonates. Of course, a combination of positive and negative examples also may serve one's needs, as long as the message is clear and concise.

Another matter to consider is convincing managers that the idea is not only compelling but also applicable and affordable. The positive attitude almost always includes such proof, as it is quite simple to demonstrate how applicable and affordable the management of lessons and good practices can be using a real-life example (or at least how easy and cost effective the management of an easily accessible knowledgebase can be).

Scope

Once management has approved the decision to run a life-cycle model of lessons and good practices management, two more important decisions must be made: scope and order. Many important management ideas presented to the organization are most relevant to one or two specific groups or units. These naturally will define the scope. Lessons management, however, is relevant to all divisions, units, groups, and teams in every organization. It is relevant of course to all those in charge of core activities (R&D, marketing, sales, and service), yet it is equally relevant to others as well, including finance, IT, HR, and operations. There is not a single group or division that will *not* gain from this chain of events.

Because of this, it is tempting to start the program broadly by applying it to all divisions and units. Why wait and work with only a small part of the organization?

The recommendation is not to wait; rather it is to work step by step. Initially work with only two pilots, and gradually add in others. Starting off with a pilot and enlarging scope gradually offers three major advantages. First, the methodology described in this book is general and as such it can, and even should, be tailored to suit each specific organization. Second, implementing the program, especially changing people's habits (cultural change) requires resources (time, management attention, and budget). Effective implementation requires staging. And, third, people find it easier to follow success. Grading the implementation may encourage communicating other stories of success, thus easing the entrance and cooperation of the next groups that decide to implement this methodology.

Choosing with Whom to Start

With which groups is it best to initiate the process? Which are the most suitable groups to take part in such a pilot of life-cycle model of lessons and good practices management?

Usually, three key factors should be considered when choosing the pilot groups: need, chances of success, and looking forward.

Need deals with the groups that have the largest gap between actual performance and the *ideal*, required performance. If a group is considered excellent, it can always continue improving, and it probably will be easy to implement these new methods of work within the group. Yet their level of anticipated improvement cannot serve as an apt pilot for applying this method within a group that has significant gaps between their "ought to" and their "is" reality.

Although need is the most important factor when choosing a group we should consider two additional factors: chances of success and looking forward. The "chances of success" factor has to do with the group's ability to institutionalize the life-cycle model of lessons and good practices management. If the group is too busy with the here and now, they will find it hard to complete such a move, and if no leader emerges or no one is assigned as project manager, for example, they will have an equally hard time implementing the change. In later stages, we may have to work with such units that require these solutions and yet are less cooperative. Nevertheless, in the pilot stage, it is allowed, and even recommended, to work with a group that has a higher chance of success.

The "looking forward" factor is also used in pilot stages. Once the pilot groups complete the move and are satisfied, they should be asked how easy it will be to prompt the organization to implement this process more broadly. This has to do with the ability of other groups to identify with the pilot groups and think, "If it worked for those guys, no reason it shouldn't work for us as well!" This factor, unlike the two others, is somewhat external to the estimated success of the pilot.

Instead, it regards the pilot as a first stage to be leveraged and then questions its ability to do so successfully.

Determining the Order in Which to Implement Change

To all those who were patient enough to read through the entire chapter, this last section is probably the most surprising and least predictable. It is nevertheless one of the more important sections, so I urge you to read through it as well. Moving to a life-cycle model of lessons and good practices management within an organization or even some specific units is a big change. Some will find it easier than others, as they already have parts of the puzzle in place and must alter only a limited number of pieces. Others, however, have a big change to digest all the new information, and as any hard-to-swallow matter, it should be broken down to smaller components. So here comes the surprise.

If you have to implement it all, do not start from the beginning; rather, start from the middle.

What does this provocative assertion even mean?

The life-cycle model of lessons and good practices management has been described as including several stages: creating the knowledge (debriefing, learning from quality processes, and learning from experiences); managing its outcomes (actions, practices, and lessons); embedding the knowledge into the organizational environment; and helping people request the knowledge before the next action.

So where should one start? Field evidence has shown that it is best to start from managing a knowledgebase. Existing organizations already utilize bodies of knowledge. The best way to start is by capturing this knowledge and building a knowledgebase based on this knowledge. If the organization holds documentation, including possible lessons or practices, one can use it. Otherwise, interviewing subject matter experts as to their accumulated knowledge is a suitable solution. Usually, parts of the knowledge are explicit and easy to share, whereas others may be tacit. Start with the experts easiest to interview; continue on to the others, showing them what already has been collected. Most people find it easier to collaborate if presented with concrete examples. The more lessons and practices you have listed, the easier it will be to collect the next ones.

Only after a substantial portion of lessons and practices resides in the knowledgebase (between 50 and 100 lessons, depending on the subject), and after these have been purified (as explained in Chapter 8), it is safe to continue and analyze how this new knowledge can be embedded into the organizational environment (Chapter 9). At this stage, the knowledgebase can be presented, communicated, and suggested to users when necessary, teaching and helping workers request the knowledge before action. By doing this, we are providing our workers with a vital tool that will assist them in best performing their tasks and making decisions.

After all these stages are completed, and the workers indeed sense the benefits of using this knowledge, then and only then should they be instructed to debrief as well. This will, along with additional experiences and knowledge captured from quality-based processes, create the new organizational knowledge.

Why do we recommend tackling this in such a peculiar order? Why start from the middle and not from the first stage? For one reason: it is easier to complete the implementation process this way. Starting from the middle shows people what they get, before asking them to invest time into the process. They use the lessons learned, before they are even asked to create new ones. We cannot always afford to do so, but regarding managing lessons learned and best practices, it is possible, and even simple, to follow this approach.

Summary

We are nearing the end of the book. We jumped into the water; we even taught people how to swim. We now have to offer a lifebelt and other swimming aids that can be of assistance until they really know how to stay afloat. The next chapter deals with these temporary aids, specifically it deals with change management. Organizations deal with this issue no matter what change is initiated. Yet it also is required in this context, and so a special chapter is devoted to this important issue. I hope it indeed will serve its purpose, which is to provide ideas for successfully changing people's habits and turning the life-cycle model of lessons and good practices management into an integral part of the organizational DNA.

Reference

Milton, N. and Lambe, P. *The Knowledge Manager's Handbook: A Step-by-Step Guide to Embedding Effective Knowledge Management in Your Organization.* Philadelphia, PA: Kogan Page Publishers, 2016.

11
THE CULTURAL CHANGE

Let us look at a wonderful cartoon about change (source unknown).

Who wants change?

Here comes the punch line:

Who wants to change?

The Goldratt Research Lab conducted research in 2011 regarding the success of change initiatives in organizations during the past 30 years. Our world constantly is changing; technology changes as do we. Yet the success of initiated changes, according to Goldratt, has not changed over the years. It was only 30% in 1980, and remained the same in samples taken every 5 years, up until 2010. The 30%

success rate means 70% failure! This means that most of the decisions management makes and believes are good for the company are not actualized, as the desired change is not completed.

John Kotter, one of the known thought leaders regarding change management, published the book *Leading Change* (1995) in which he defined eight steps to follow to manage this process. Many organizations learned his methodology; many organizational and business consultants, including Kotter himself, have earned (at least) part of their living from these methodologies. And yet as Goldratt's research has shown (and perhaps from additional sources) change management has not become easier; organizations consistently fail in an astounding 70% of the cases. Nevertheless, Kotter continued spreading his ideas. A consultant named Holger Rathgeber suggested that although the methodology seemed easy to apply, it was rather challenging to understand and deploy, and suggested that it might be easier to learn via a parable. He and Kotter came up with a new book coupled with a series of demonstrational workshops, telling the story of a penguin trying to lead a change in some penguin colony in Antarctica, going through the eight stages of change previously defined by Kotter (Kotter and Rathgeber, 2005). Unsurprisingly, the statistics did not change. Kotter, however, did not give up. He formulated a survey, asking organizations that had tried to follow his suggested eight-step methodology to identify the point at which they failed. That is, at what stage does the change process halt and eventually disintegrate? More than 250 organizations participated in his survey and the results were astonishing. Most organizations got stuck at stage one! Most organizations that decided to implement a change, and even decided to invest resources in the process, did not even succeed in reaching the second of eight stages. And what is Kotter's first stage that so many organizations do not surpass? Generating a sense of urgency—realizing that they must change immediately and cannot postpone this change to any near or far future event (Kotter, 2008).

Over the years, we have been exposed to a great deal of theories concerning both organizational and individual resistance to change. Individual models can be categorized into four types of approaches: behavioral approaches (i.e., Watzelweek, Weekland, and Fish); cognitive approaches (i.e., Beck); psychodynamic approaches (i.e., Kübler-Ross and Satir); and humanistic approaches (i.e., Maslow). Organizational approaches are even more complicated. Theorists explain their models through a series of allegories, presenting organizations as machines, political systems, organisms, or ever-changing flows (Levy, 2015). Unfortunately, change has not become easy as a result of studying these models.

Returning to Kotter, we still have cause to be optimistic. During the past few years, a handful of leading thinkers have developed new methodologies tailored to handling change management processes in organizations. Not one of them claims to have come up with a recipe that can be applied to all needs, or even to one specific organization and situation, as is, with no alterations. It also seems that these leading thinkers learn each other's ideas and do not try to reinvent the wheel. Rather, they "sample" parts of other methodologies and then add their inputs about practical recommendations for dealing with this unacceptable situation of such high rates of failure.

In the context of this book, cultural change should be implemented through two processes. First, we must change our workers' organizational habits, so that the organization embraces the life-cycle model of lessons and good practices management. Second, once such a life cycle is adopted and new lessons and practices are created, a process is required to implement the recommendations these yield and to continually improve the organization's performance toward excellence.

Before embarking on such a move, the organization wishing to implement changes must understand that these changes cannot be implemented simply by sending out orders, instructions, or explanatory memos. To change people's habits, an entire process should take place; in many cases, the required process is not a short one (a sprint); rather, it is a gradual process that may require investing time and effort (a marathon). This process must be managed as a project, complete with goals, activities, and milestones, and as such, it must be defined and adapted according to the organization and situation in which it is applied.

Following is a series of good practices that can be helpful when completing such a change management program within the boundaries of reasonable effort. We must keep in mind that if the change is not applicable, it better not be introduced in the first place. These recommendations are based on ideas developed by leading thinkers considering twenty-first century employees and their typical characteristics. I present only those ideas that I have found to be most productive in these cases.

Initiating the Move

In 1962, Kurt Lewin defined three steps to be considered when performing any change: unfreeze, change, and refreeze. Decades later, Michael Fullan, a leading thinker in the field of change management, addresses the issue of change management in his book *The Principal* (2014), describing the future of this profession. Fullan, based on a model formulated by Lyle Kirtman, suggests a process that includes seven components and begins with an advanced version of Lewin's "unfreeze" stage. This component deals with challenging the status quo. One problem we face when attempting to initiate any change (specifically changes related to new lessons) is that people are overwhelmed with information and each piece of data potentially demands change. We cannot implement all the required changes derived from this overload of information; sometimes, we cannot even spare the time to process all this data. We, therefore, shelter ourselves from this flood of information by blocking our minds, not really paying attention to any of the suggestions. For something to pass the filter of our attention, it has to either come from someone we highly respect, or we have to be shaken into attention. The idea is to crack our mental "wall of protection" and enable the desired information to flow through the cracks we create. That is the rationale behind the "unfreeze" and "challenging the status quo" stages: cracking open our cocoon of convenience, motivating us to seek change, and a willingness to listen to the message.

Adding Kotter's sense of urgency to this recommendation ensures it has a good chance of happening: cause the earth to shake, so people realize they want or even need to change; provide them with a sense of urgency, so that they want to do it now rather than endlessly procrastinating.

Helping People through the Journey of Change

Unlike a wheel that can potentially turn forever, when dealing with change management, people and organizations require assistance to continue changing existing habits and routines even after somebody gets things moving. It takes much more than convincing them that a change has to occur and has to occur now.

A well-known phrase regarding decisions says that 80% of any job can be credited to an emotional decision with the remaining 20% being dedicated to coming up with logical reasons (one might say, excuses) to justify our (emotional) decision.

The same phenomenon occurs with change. We can be convinced that something should be changed (e.g., our weight), yet we will find it hard to follow through with the required change. This is not solely for objective reasons. Sometimes, our mind takes us back, again and again, to square one.

Two tools can help us to move this stuck wagon forward.

The first tool is one of the "change program" components suggested by the Heath brothers. These two brothers, Chip, a professor of Organizational Behavior in the Graduate School of Business at Stanford University, and Dan, a senior fellow at Duke University's CASE Center, have coauthored several successful books dealing with management in organizations. Their ideas are refreshing and occasionally extraordinary. A change implementation process they suggest (Heath and Heath, 2010) is composed of three main components: the rational component (which they refer to as "directing the rider"); the emotional component ("motivating the elephant"); and, finally, an environmental component ("shaping the path"). In some cases, a change can be completed when the process leads us to work in a new way. Chapter 9 of this book, which suggests embedding the knowledge into the organizational environment, is based on this same rationale. If someone opens an e-form and it contains a hyperlink to lessons, people naturally will use it; they might not even be aware of the organizational change. The Heath brothers use a fabulous example regarding nurses working in a hospital failing in some cases to give patients the correct medicine. Searching for the reason for these recurring mistakes, they found that the nurses were disrupted by physicians or patients who requested they do something else, distracting them and causing them to err. Change implementation is difficult in this case because of all three parties: nurses, physicians, and patients. Hospital management decided to change the environment rather than attempt to convince people that lives would be saved if medicine could be provided with no mistakes. Special jackets and caps were provided to the nurses, to be used only when dealing with medicines handed out to the ill. These jackets had instructions printed on them asking people not to talk to the nurses wearing these jackets, as they were distributing medicine. Surprisingly, mortality levels dropped by more than 50% as a result of this change. No rational explanations were provided, nor were any alarming or emotional means used to raise awareness. An environmental change can sometimes help us lead an organization toward change, and in our case, to implementing a "life-cycle model of lessons and good practices management" program.

One way to help a change become a habit is by shaping the path, thus allowing us to smoothly adjust the change. It seems great; no sweat. Yet this solution has one small disadvantage: it is not always applicable. In too many cases, no shortcuts can be taken; we have to do things the old way.

This brings us to our second tool. The term "tool" is quite appropriate in this case, as this may seem to be a well-structured methodology that one logically can follow and use to bring about change. This tool, or methodology, is not the only structured one, yet many organizations find it to be the most practical one—that is, a methodology that indeed works. It was developed by Jeff Hiatt, an engineer and program manager for Bell Labs who founded the Prosci consultancy group started in 1994. The tool is based on research that was tested on 1600 project leaders and teams (Hiatt, 2006). The methodology in question was designed to assist change on an individual level, keeping in mind that organizations are composed of individuals and change has to occur on both levels—organizational and personal. The methodology is based on five components: awareness, desire, knowledge, ability, and reinforcement (ADKAR).

It is suggested to implement the ADKAR model, take its ideas, and push them even further: ADKAR can serve as a way of implementing change not only for individuals, but also for teams, departments, divisions, or even an entire organization. During every stage of change implementation, the five components may serve as potential hindering factors, slowing or preventing the change:

- People may not be aware of the change or of its necessity;
- People may not wish or agree to change;
- People may not know what the change includes or how they are expected to act on it;
- People may not possess the abilities and capabilities required to actually change old work habits; and
- People may lack reinforcement gained by repeatedly performing the new routine, embedding the change into each worker's individual work DNA as well as that of the organization as a whole.

So, you might ask, what actual practical suggestion do you offer? How can these ideas be transformed to practice, driving the change?

Whichever organizational unit is chosen for change, it should be analyzed, whether by survey or any other suitable, reliable analysis technique, to discover the one or two main hindering factors currently affecting the unit. Organizations may regard the entire organization as a single unit, or may opt to analyze these factors separately as subunits, designing a change program for each according to its current state in terms of change implementation. One may wonder: *Why focus on only one or two factors, while in many cases all five main factors are hindering the change?* Focus is important, as working in too many directions at each point in the process may not be effective. In situations in which knowledge is missing, start with the first two components (awareness and desire), and then continue the next week or month to the next two stages (desire and knowledge), and so on. The chronological order in which these stages are pursued is important because these five components represent five milestones to be reached, and they depend on each other. If we are not aware of the change, how can we desire it?

When the organization knows which of the hindering factors it is currently focusing on, it is easier to use the regular communication channels and other known techniques (including those still in development) to promote an effective change. For example, if the organization decides it needs to work on its "desire" component, posters and newsletters should focus on why said change is beneficial. This is the time when "hard" and "soft" rewards should be touted. When the organization decides to focus on work-habit reinforcement, it might hold competitions rewarding those who achieve the desired behavioral change. People will grow accustomed to working as expected. As mentioned, these are only select examples and ideas; the list goes on and on.

Summary

One of the most frustrating things about lessons learned and good practices is that occasionally we know how to improve, and even know how to use a life-cycle model of lessons and good practices management, yet when attempting to apply this knowledge, we encounter an organization that seems to resist change. They refuse to implement useful knowledge that certainly contributes added value, competitiveness, and ultimately leads to success.

This chapter offered recommendations about how to implement such change and turn the life-cycle model of lessons and good practices management into reality.

We still have one issue yet to handle: defining the roles that must be filled to implement all the ideas suggested in this book. That is the subject of the next chapter.

References

Fullan, M. *The Principal: Three Keys to Maximizing Impact*. San Francisco, CA: John Willey & Sons, 2014.
Heath, C. and Heath, D. *Switch: How to Change Things When Change Is Hard*. New York: Broadway Books, 2010.
Hiatt, J. *ADKAR: A Model for Change in Business, Government, and Our Community*. Loveland, CO: Prosci Learning Center Publications, 2006.
Kotter, J.P. *Leading Change: Why Transformation Efforts Fail*. Boston, MA: Harvard Business Press, 1995.
Kotter, J.P. *A Sense of Urgency*. Boston, MA: Harvard Business Press, 2008.
Kotter, J.P. and Rathgeber, H. *Our Iceberg Is Melting: Changing and Succeeding under Any Conditions*. New York: Macmillan, 2005.
Levy, M. (2015). If Only We Build KM Solutions, Users Will Come. Or Would They? www.kmrom.com (accessed August 16, 2016).

12
ROLES IN THE ORGANIZATION

A tree, any tree, is composed of many parts; it has green leaves, responsible for reproduction, and it has roots that are in charge of sustenance and ensuring that the tree is supplied with all it needs for optimal growth. There are also branches, tissue, a trunk, possibly flowers and fruit, and many additional components, each playing a role in turning the tree into what we consider a tree.

Surprisingly, a tree has no head; no leader or manager in charge of orchestrating the different components, telling them what to do and how to do it or monitoring their activities.

Trees are not an exceptional headless model. The Internet, like many other compound systems, does not have any one manager in charge.

So, when we come to manage the lessons in our organization, can we settle for a well-defined process and no lessons manager? Can we manage our knowledge in a compound system, in which everyone is in charge of debriefing; everyone inserts their lessons into a shared knowledgebase that everyone uses whenever suitable, with no additional management?

Anyone who has spent several years working in organizations knows that this suggested method probably would not work in the majority of organizations.

Several processes are too complicated for us to ask the employee in charge of debriefing to be responsible for performing these duties as part of the knowledge creation process. Furthermore, cleansing processes should be performed routinely, and these tasks do not fall under the natural role of anyone responsible for debriefing.

We also have tasks that deal with communicating the need for lesson management, showing those who are supposed to manage their lessons how to optimally perform; and we have tasks dealing with change management, helping the organization change its habits and actually manage its lessons, experience, and best practices as part of its ongoing processes and culture.

Some of these roles were discussed in Chapters 6 through 8, dealing with processes regarding knowledge created and methods to encourage its use, as well as in Chapters 10 and 11, dealing with organizational change. This chapter focuses on all processes, attempting to suggest a method in this madness and developing better understanding of what and who should be involved as organizations decide to leverage their lessons management.

Five roles are necessary to build and sustain the life-cycle model of lessons and good practices management, as defined. These include the following:

- sponsor
- lessons manager
- lessons knowledgebase manager
- subject matter experts
- employees

The Sponsor

No news here. As in many other organizational issues, leadership is required, and the sponsor serves as the voice of organizational management. The sponsor is required to demonstrate leadership and commitment, both in suggesting a system of a full life-cycle model of lessons and good practices management and in ensuring its optimal conduct so that, in turn, it can benefit the organization and its continuous learning and success. The International Organization for Standardization management standards refer to several aspects of this role (ISO, 2015), including the following:

- ensuring alignment between the defined objectives of the lessons learned and those of the organization;
- ensuring integration of the lessons into the organization's business processes;
- ensuring resources as required;
- communicating the importance of effective lessons and good practices learning in its new full life-cycle definition;
- ensuring the new way of work achieves its defined outcomes; and
- directing people to perform their roles as defined in this chapter.

Organizations know how to define a sponsor and his or her roles. Not much has to be said, besides this one piece of advice: do not risk the idea; do not try skipping this role.

The Lessons Manager

Organizations that decide to implement an advanced methodology of lessons management should appoint someone to be the lessons master. Following are the responsibilities of this role:

- assisting in first or complicated debriefings;
- training units about how to build a knowledgebase, purify lessons and good practices, and find ways to embed these in the organizational and business environment; and
- initiating cross-unit debriefs and serving as project manager in the first stages of the implementation of this methodology in the organization.

These organizational functions have no justification or purpose if isolated. Without specific units, or cross-organizational disciplines of interest, it is more like fiercely pressing the gas pedal while the car is in a neutral mode.

The following roles shall be defined for each specific knowledgebase built. They may exist across units; for example, a project's lessons and good practice knowledgebase; it also could be specific to engineering personnel (including design ideas, manufacturing, and safety). Knowledgebases should not be overly focused, but there probably are not too many good reasons to file safety and marketing issues together.

The Knowledgebase Manager

The knowledgebase manager deals with the discussed discipline. Two types of individuals found in organizations can serve as knowledgebase managers: an expert or a facilitator. Following are their respective profiles.

Expert profile: the chosen knowledgebase manager is the number one expert, or one of the leading experts, of the discipline in the organization or unit (if implemented in a specific one). This expert is in charge of the following:

- designing the structure of the knowledgebase and initially manage its population with relevant valuable knowledge;
- encouraging lesson-learning sessions and debriefings regarding the subject at hand;
- individually purifying lessons and good practices received and adding them into the knowledgebase;
- suggesting means by which the lessons and good practices will be embedded in the organizational and business environment;
- directing people to use the knowledgebase;
- analyzing and learning how to nurture usage; and
- occasionally cleansing the knowledgebase, verifying its content is still relevant.

In some situations, such a profile does not fit the organization's needs, especially when the knowledgebase comprises several areas of expertise (and would each need its own expert) and yet serves as one unified knowledgebase. One such example is a hazard-safety knowledgebase: the knowledge deals with a large variety of expertise, such as electricity, toxins (relevant to labs), working at heights, ergonomic issues, and construction contractors. Relying on one expert to purify and validate all lessons and good practices may result in a less professional knowledgebase.

The knowledgebase manager will serve in these cases as a facilitator, receiving lessons and good practices and directing them to a specific expert. The facilitator will have a similar role to those noted, yet some roles will be delegated to others, ensuring that others do the job as required.

Life is always more interesting in practice than in theory, and some organizations have combinations of these two roles. For example, some large pharmaceutical R&D departments scattered over seven countries, appointed seven experts with the same areas of expertise, one representing each country, and all deciding together on a well-defined methodology as to every new best practice. In these cases, the knowledgebase manager is a member of the group and is in charge of verifying that decisions indeed are taken and that practices are checked.

Subject Matter Experts

When knowledgebases are managed by experts covering different areas of expertise (the facilitator model), the term subject matter experts is used to identify these various experts and emphasizes their role.

Employees

Last, but certainly not least, any organization wishing to implement a full life-cycle model of lessons and good practices should regard the employees who are to use the knowledge with respect. One can construct methodologies, set knowledgebases, and populate them with content, but if the employees will not use the new knowledge, no learning will take place.

Employees are required to suggest new good practices for the knowledgebase. It now is understood that wisdom comes from among the crowds, and designing a knowledgebase to include only the wisdom of a small set of people will not encompass the knowledge of the employees who are actually working—maybe succeeding, maybe failing, but for sure, accumulating new experience and knowledge along this journey.

Employees are required to use the knowledge. With the overflow of information and knowledge, in many cases people do not remember using the knowledge they have learned themselves, much less those lessons learned by others. Accessing a focused knowledgebase consisting of several hundreds of lessons and practices, and having an easy way to access specifically relevant knowledge regarding the current situation and context, can help each and every one perform their tasks better, making more effective decisions when reaching important milestones.

Where does this bring us? Simply put, to the end of this book and to a summary of the main ideas presented throughout our journey.

Reference

ISO. *Annex SL: Proposals for Management System Standards*. Geneva, Switzerland: International Organization for Standardization, 2015.

13
Summary

Every year, on October 13, the Finns celebrate their international day dedicated to... failure.

They encourage people to dare to fail, share their shortcomings with the world, and learn from them. They even maintain a special website for this occasion that features explanations, quotes, and instructions (see dayforfailure.com). What are they trying to achieve? Is this their unique way of telling people that all ways are equally legitimate, meaning even failure is a legitimate route?

I reckon that a deeper concept lies behind this initiative. To improve, one has to first experience and then learn from this experience. Calling for failure is a way to invite people to dare and experience new things, thus creating new opportunities for learning. In this book, I have offered a perception that focuses on the second part of this equation. Once people experience an interesting situation—whether a success or a failure—what can be done to ensure that this new knowledge actually is used when relevant in the future and that organizations learn from and improve their business conduct?

Surprises

One of the first ideas presented in this book dealt with the question of when to debrief. This question is relevant for organizations as well as for individuals. It is common to debrief on well-defined occasions either based on routine (after a project or bid) or other clearly defined occurrences (great loss or damage). An analogy to this idea may be deciding when to call our grandparents; I might decide, at some point in my life, to call them every second weekend and on holidays. These are well-defined occasions. I also might decide to call them every time I experience something that reminds me of my childhood or of them. Such phone calls do not substitute the routine ones that regularly ensure the continuous relationship. They are an adage that contributes to the relationship's authenticity.

A similar idea (presented in Chapter 3) was to debrief whenever we are surprised. The concept of surprises is relevant to both positive experiences as well as negative ones. It gives us a good starting point, an anchor with which to ground our learning. Just like the nostalgic memory that can serve both as a trigger for conversation and as the content of some routine call, a surprise may trigger the process of learning yet also serve in a routine debriefing as a way to focus on the learning despite various distractions.

Debriefing Techniques

Chapter 4 reviewed two of the many possible debriefing techniques. No one debriefing method is absolution; and no single technique is superior. Every organization can and should choose a methodology that suits its needs and nature. Not only can it choose from among the existing methodologies presented in this book (or in other textbooks), but the organization can also develop its own methodology.

Two recommendations should be taken into consideration when selecting or developing a debriefing methodology:

1. Whatever methodology the organization chooses, it should be one that incorporates truly understanding the case's details and causes.
2. The organization should ensure that the chosen methodology focuses on the future; specifically on lessons that can be learned to improve future organizational performance.

In short, organizations should base their learning from the past by logically analyzing it, thus turning it into a better potential future.

Additional Sources

Debriefing is a process that enables us to develop new knowledge and learn. Yet debriefing is not the only process that serves as a source of new organizational knowledge. Along with debriefing processes, two additional processes were presented; these processes can serve as sources of new knowledge that share the same structure that lessons possess. These include quality-based processes (described in Chapter 5) and experiencing (described in Chapter 6). These two processes occur in our personal and organizational lives, and it is a waste not to use their outcomes as sources of new knowledge. Furthermore, in many cases, significantly less energy is required to collect this additional knowledge than that invested in debriefing processes. In most relevant situations, no one feels threatened by the process and the quality of the additional knowledge has high potential. Examples of quality-based processes included Plan-Do-Check-Act cycles (PDCAs), quality audits, and customer feedbacks. The term *experiences* include all life or work experiences. We learn from almost whatever we experience, whether consciously or unconsciously.

Outcomes

We are accustomed to thinking about lessons in terms of actions (i.e., immediate tasks to be completed by an assigned person by a certain deadline). Life, however, is more complicated. Some outcomes are filtered because when reviewed using a holistic perspective we realize they were not as important as originally considered, or they may even have turned out to be faulty. Following the filtering process, more ideas for improvement usually emerge than any organization can actually implement. In too many a case, after committees complete their work or debriefing sessions take place, organizations are overwhelmed with suggestions.

Furthermore, some lessons are specific and others are general; some lessons are definite and others are merely recommendations.

The idea presented was to separate the new knowledge into three types:

- Actions (described in Chapter 7) representing well-defined tasks; these usually can be defined by a timeline and an individual responsible to verify completion.
- Lessons and good practices (described in Chapter 8), including the remaining new ideas that were selected from which to learn, yet were not integrated via actions or changes. They may represent recommendations to be considered by the manager before making a decision toward some action, or they may represent best practices to be embedded in specific situations.

- Changes (described in Chapter 12) representing major lessons to be integrated into routines. It is recommended to choose and prioritize only a few changes to improve the chances of truly embedding and assimilating these changes into organizational or personal DNA.

An important idea presented in the book is that these should be handled as separate items, not as part of some debriefing or other document. They should be purified: worded, generalized, and merged with other existing lessons and good practices. They should be written as bottom lines, easily accessible thanks to categorization or a simple search.

Implementing these lessons leverages usage and is in itself a major step toward reuse and business improvement.

Closing the Loop

Although actions, changes, and lessons upgrade the way most organizations handle their knowledge, additional steps can be taken to further increase our chances of using this new knowledge and ultimately improving.

Two main additional ideas were presented. The first idea (Chapter 9) dealt with embedding the new knowledge into a life environment, easing its usage and, in some cases, even making it more difficult to err again. Embedding can take place in training, in forms, in meetings, in computing systems, and in many other existing mechanisms. The second idea (Chapter 10) dealt with proactive ways to raise awareness about these lessons, notifying people about the availability of this new knowledge. These two steps close the loop forming a full life-cycle management of lessons and good practices (Figure 13.1).

SUMMARY

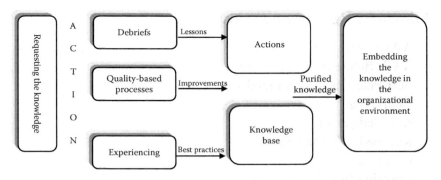

Figure 13.1 The lessons and practices cycle—the full cycle.

Making such a change and implementing the full life cycle is a journey, as has been described throughout this book.

Just remember that even the longest journey starts with one step. In this case, if implemented wisely, low hanging fruit after each step is taken are just waiting to be picked.

Bon appétit.

Glossary: Review of Terms

AAR: After Action Review
BAR: Before Action Review
Best Practices (Good Practices)
Debriefing
Knowledgebase
Knowledge Management
Learning
Learning from Experience
Learning from Quality-Based Processes
Learning Organization
Lessons/Lessons Learned
Team Learning

AAR: After Action Review

There are many debriefing techniques; one popular method, originated in the U.S. Army, is the after action review (AAR) technique.

AAR asks four questions:

1. What were our expectations?
2. What happened?
3. How can we explain the unexpected difference between the two?
4. What do we recommend?

The U.S. Army, in its "A Leaders Guide to After-Action Reviews," defines AAR as "a professional discussion of an event, focused on performance standards, that enables soldiers to discover for themselves what happened, why it happened, and how to sustain strengths and improve on weaknesses" (1993, p. 4).

BAR: Before Action Review

The BAR is a well-known technique that is complementary to the AAR, in which participants assess processes and their results before actually launching the project. This is done to approach the project and its risks as prepared as possible.

The BAR technique asks the following questions:

1. What is expected?
2. What can go wrong?
3. What can we learn from the past?
4. What do we recommend?

Best Practices (Good Practices)

Best practices are recommendations for future behavior and comprehension based on past events and learned from experience.

Weber, Aha, and Becerra-Fernandez define best practices as follows:

> Descriptions of previously successful ideas that are applicable to organizational processes. They usually emerge from reengineered generic processes
>
> *(O'Leary, 1999, cited in Weber et al., 2001, p. 1)*

> They differ from lessons in that they capture only successful stories, are not necessarily derived from specific experiences, and they are intended to tailor entire organizational strategies
>
> *(Weber et al., 2001, p. 1)*

Now, some organizations refer to "good practices," as we cannot always assure the practices to be the "best."

Debriefing

A process in which an individual or group analyzes an activity that occurred and sets recommendations for future behavior based on that analysis.

Markullis, in her article "A Brief on Debriefing: What It Is and What It Isn't" (2003), argues that there is no consensus regarding the definition of debriefing and reviews several definitions. Nevertheless, Markullis integrates all definitions into the following formula: "Learning through reflection on a simulation experience" (Markullis, 2003, p. 177).

Jarvis defines debriefing as reviewing past events with the objective of gaining insights from mistakes and successes that will improve future decision-making. By debriefing, we are attempting to translate tacit lessons from the highly subjective and personal experience to explicit terms, so that these lessons can become available to organizational knowledge management processes (Jarvis, 1999).

Knowledgebase

A knowledgebase is a database that mainly stores information and knowledge.

Wikipedia defines the knowledgebase as a "technology used to store complex structured and unstructured information used by a computer system. The initial use of the term was in connection with expert systems which were the first knowledge-based systems... The term 'knowledge-base' was coined to distinguish this form of knowledge store from the more common and widely used term database."

The term is commonly used for several types of content, among them lessons learned.

Knowledge Management

Knowledge Management is a combination of processes, actions, methodologies, and solutions that enables retention, sharing, accessibility, and development of organizational knowledge.
Knowledge management is defined by Raitt, Loekken, Scholz, Steiner, and Secchi as follows:

> Knowledge Management is a discipline that promotes an integrated approach to identifying, managing, sharing, and leveraging all of an enterprise's knowledge and information assets, by continuously employing a set of policies, organizational structures, procedures, applications, and technologies. These knowledge and information assets, often referred to as the "corporate memory," include databases, documents, policies, and procedures (i.e., "explicit" knowledge), as well as previously unarticulated experience and expertise resident in individual workers' brains (i.e., "tacit" knowledge). Knowledge management thus aims at

leveraging the ability of the capable, responsible, autonomous individual to act quickly and effectively

(Raitt et al., 1997, p. 112)

Learning

Learning is an act of knowledge acquisition followed by implementation, dissemination, and results. Experts believe that learning implies behavioral change of the learner as a direct result of the possession of new knowledge.

"Learning can be defined as changes in behavior resulting from experience" (Bunning, 1992, p. 7). Jarvis refers to earlier sources (English and English, 1958): "the traditional definition of learning in psychology literature is a shift in performance when the stimulus situation and motivation remains essentially the same" (Jarvis, 1999).

Hence, learning involves much more than acquiring knowledge; it implies implementation, dissemination, and results.

Learning from Experience

Learning from experience is essentially a subconscious process. The learning is performed without any proactive session, such as debriefing or quality-based processes.

According to Dirkx and Lavin:

> Experience-base[d] learning is a concept and a phenomenon which represents the core of the research-to-practice issue. As a phenomenon, the term refers to the fact that learning takes place within the crucible of our life experiences and cannot be separated from them. As a concept, experience-based learning provides a means of developing a theoretical understanding of how lived experiences influence what is learned and vice-versa
>
> *(Dirkx and Lavin, 1991, p. 7)*

Cell refers to two levels, primary thinking and secondary reflection:

> As we are transacting with our world our minds are continually at work interpreting these transactions and the situations in which

we enact them. This is primary thinking. It is the spontaneous and usually habitual activity that forms the background of our actions and reactions. We are vaguely aware of the same parts of it and not all aware of others. Much of it, that is, takes place subconsciously. Then at times we disengage ourselves from our involvements to think about some of them more carefully and systematically. This is secondary reflection

(Cell, 1984, p. viii)

Learning from Quality-Based Processes

Learning that is performed as part of the implementation of quality methods, such as Root Core Analysis (RCA) or PDCA.

Learning Organization

An organization's ability to continually improve based on learning, as defined by Senge:

> An organization that is continually expanding its capacity to create its future. For such an organization, it is not enough merely to survive. "Survival learning" or what is more often termed "adaptive learning" is important; indeed, it is necessary. But for a learning organization "adaptive learning" should be joined to "generative learning," learning that enhances our capacity to create
>
> *(Senge, 1990)*

Senge also describes the change as a "metanoia"; a mental shift that organizations experience when becoming a learning organization.

Lessons/Lessons Learned

Lessons learned are recommendations for future behavior and comprehension based on past events and learned through debriefing processes.

Lessons and lessons learned are synonymous and are used interchangeably in knowledge management literature.

Weber and Aha define lessons learned as follows:

[K]nowledge artifacts that convey experiential knowledge that is applicable to a task, decision, or process such that, when re-used, this knowledge positively impacts an organization's results

(Weber and Aha, 2003)

Team Learning

Shared learning by a group of people, as defined by Senge:

Team learning is the process of aligning and developing the capacity of the team to create the results its members truly desire

(Senge, 2006)

References

Bunning, C. Turning experience into learning: The strategic challenge for individuals and organizations. *Journal of European Industrial Training, Bradforn* 16(6): 7–12, 1992.

Cell, E. *Learning to Learn from Experience*. Albany, NY: State University of New York, 1984.

Dirkx, J.M. and Lavin, R. Understanding and Facilitating Experience-Based Learning in Adult Education: The FOURthought Model. Paper presented at the Midwest Research-to-Practice Conference, St. Paul, MN, October 1991. https://msu.edu/~dirkx/EBLRVS.91.pdf (accessed March 05, 2017).

Jarvis, C.B. Learning (or Failing to Learn) from Experience: The Dysfunctional Implications of Counterfactual Thinking in Reviewing Past Events to Improve Marketing Performance. PhD diss., Indiana University, 1999.

Markullis, P.M. A brief on debriefing: What it is and what it isn't. *Developments in Business Simulation and Experiential Learning* 30: 177–184, 2003.

Raitt, D., Loekken, S., Scholz, J., Steiner, H., and Secchi, P. Corporate knowledge management and related initiatives at ESA. *ESA Bulletin* 92: 112–118, 1997.

Senge, P.M. *The Fifth Discipline*. New York: Random house, 1990.

Senge, P.M. *The Fifth Discipline*, 2nd ed. New York: Random house, 2006.

U.S. Army. (1993). TC25-20: A Leader's Guide to After-Action Reviews. http://www.au.af.mil/au/awc/awcgate/army/tc_25-20/tc25-20.pdf (accessed March 05, 2017).

Weber, R.O. and Aha, D. Intelligent delivery of military lessons learned. *Decision Support Systems* 34(3): 287–304, 2003.

Weber, R., Aha, D., and Becerra-Fernandez, I. Intelligent lessons learned systems. *Expert Systems with Applications* 17, 2001.

Wikipedia. Knowledge Base. https://en.wikipedia.org/wiki/Knowledge_base (accessed March 05, 2017).

Index

A

AAR 17, 22, 26–9, 101–2, 133–4
    enlargement of 29–33
    questions in 26–7
    timing of 17–18
abstract conceptualization 49
actions 59–61, 131
active experimentation 49
ADKAR (awareness, desire, knowledge, ability, and reinforcement) 120–1
After-Action Review (AAR) *see* AAR
Aha, D. 134, 138
attachments 77
attributes 71–72
    context-based 72, 73–4
    fixed 74–7
    two-stage 73
    universal 72
audits
    quality 4, 43–4

B

BAR 101–4, 134
Becerra-Fernandez, I. 134
Before Action Review (BAR) 101–4, 134
best practices 55, 131, 134
Boiko, Bob
    Content Management Bible, The 72
British Petroleum (BP)
    Peer Assist program 100–101
Bunning, C. 136

C

Capability Maturity Model Integration (CMMI) 39
Cell, Edward 48
    Learning to Learn from Experience 10–11, 51–3, 137
change, cultural 115–22
change management 60–1, 132
    actions in 59–61
    cultural change 115–22

change management (*cont.*)
  debriefing 20
  desire to start 107
  method to start 108–10
  order of change 112–13
  pilot group for 111–12
  resistance to change 116–17
  scope of change 110–11
change program 119
checklists 88–9
CMMI 39
competition 5–6
concept-details dimension 49
concrete experience 49
context-based attributes 72, 73–4
cultural change 115–22
  assistance with 119–22
  initiating 118

D

debriefing 55–6, 97
  change, after 17
  core processes, after 17–19
  definition of 18, 34, 135
  end of project, after 17
  incorporating information gained from 9–10
  multi-case learning 33–7
  need for 16
  personnel involved in 21–3, 99
  questions for 21
  surprise, upon 18, 19–20, 129–30
  techniques 26–33, 130
  timing of 16–21
  value of 66, 72
Define, Measure, Analyze, Improve, and Control Method (DMAIC) 41–2

Deming, William Edwards 40
Deming cycle 40–1, 42, 137
Dirkx, J.M. 137
DMAIC 41–2
Drucker, Peter
  Landmarks of Tomorrow 4
  Management Challenges for the 21st Century 4
  Practice of Management, The 4

E

embedding lessons 81–93, 132
  checklists 88–9
  exercises 86
  forms 82–4
  guidelines, within 92–3
  interesting lessons session 85–6
  mistakes, reducing risk of 86–8
  online help 90–1
  procedures, within 92–3
  processes, within 91–2
  in search processes 89–90
  templates 82–4
  training 84–6
experience 47–56, 131 *see also* knowledge
  as best practice 55
  processes of learning from 48–53

F

fixed attributes 74–7
  next validation date 74–6
  sensitivity 76–7
forms 82–4
four forms of learning 48–50
four levels of experiential learning 51–3
Friedman, Thomas

INDEX

Lexus and the Olive Tree, The: Understanding Globalization 5
Fullan, Michael
   Principal, The 118

G

Gemba walk 42–3
Gembutsu 42–3
global connectivity 5–6
Goldratt Research Lab 115
good practices 55, 131, 134
Google 65, 107
guidelines 92–3

H

Heath, Chip 119
Heath, Dan 119
Hebert, Cindy
   New Edge in Knowledge, The 54
help, online 90–1
Hiatt, Jeff 120
hyperlinks
   in knowledgebase 70, 77
   to knowledgebase 83

I

information 4
   access to 7, 65–6
   core 8
   external 6–7
   internal 7–11
International Organization for Standardization (ISO) 39, 124
Internet 4, 6
ISO 39, 124
Israeli Air Force 103

J

Jarvis, C.B. 135, 136
Jerusalem Municipality 102–3

K

Kahneman, David 48, 53
   Thinking, Fast and Slow 50–1
Kirtman, Lyle 118
knowledge *see also* lessons
   access to 65–80
   availability when needed 97–104
   embedding 81–93, 132
   employee integration of 10–11, 53–6
   experience as 47–56
   external information and 6–7
   internal information and 7–11
   new, awareness of 102–3, 132
   potential 9
   retaining 62–3
   role of 3–5
knowledgebase 55, 69–73, 125
   attributes 71–2, 73–7
   benefits of 70–1
   definition of 135
   hyperlinks to 83
   lessons in 69, 77–80
   structure of 72, 73–7
   values 71–2
knowledgebase manager 125–6
Knowledge Management 135–6
knowledge workers 4
Kolb, David 53
   Experiential Learning: Experience as the Source of Learning and Development 48–50
Kotter, John
   Leading Change 116, 117, 118
   Sense of Urgency, A 20

INDEX

L

Lambe, P. 108
Lavin, R. 137
Lean Management 42
learning *see also* debriefing; experience
    accuracy in 71
    definition of 49, 136
    efficiency of 71
    from experience 47–53, 136–7
    lessons 15–23
    from quality-based processes 137
    team 138
learning from experience 47–53, 136–7
learning from quality-based processes 137
learning organization 137–8
Leistner, Frank 82
lessons 32, 131 *see also* knowledge; knowledgebase
    body of 73
    definition of 47, 138
    embedding in organizational environment 81–93, 132
    in knowledgebase 69, 77–80
    learning 15–23
    purification of 79
    values of 72
lessons manager 125
Levitin, Daniel
    Organized Mind, The 19, 87
Lewin, Kurt 60, 118
Loekken, S. 136
Lycos 65

M

Markullis, P.M.
    "Brief on Debriefing, A: What It Is and What It Isn't" 135

MCL 33–7
Milton, Nick 108
    Lessons Learned Handbook, The 98–9, 100
multi-case learning (MCL) 33–7

N

National Institutes of Standards and Technology 4

O

O'Dell, Carla
    New Edge in Knowledge, The 54
organization
    employees 127
    knowledgebase manager 125–6
    learning 137–8
    lessons manager 125
    roles in 123–7
    sponsor 124–5
    subject matter experts 127

P

PDCA 40–1, 42, 137
Plan-Do-Check-Act Model (PDCA) 40–1, 42, 137
price 3
procedures, lessons embedded in 92–3
processes, lessons embedded in 91–2
purification 79

Q

quality assurance 3–4
quality audits 4, 43–4
quality-based processes 39–46, 131
    DMAIC 41–2
    Gemba walk 42–3

learning from 137
Plan-Do-Check-Act Model
40–1, 137
quality audits 4, 43–4

R

Raitt, D. 136
Rathgeber, Holger 116
RCA 40, 137
reflective observation 49
responsive learning 52
Rogers, Everett 60
Root Core Analysis (RCA)
40, 137

S

Scholz, J. 136
search engines 7, 65
Secchi, P. 136
Senge 137–8
Shewhart, Walter 40
situational learning 52
Six Sigma 41
sponsor 124–5
Steiner, H. 136
System 1 50, 51
System 2 50, 51

T

tasks 31–2
team learning 138
templates 82–4
time 4
Total Quality Management 3
training 84–6
transcendent learning 53
trans-situational learning 53
two systems theory 50–1

U

U.S. Army 17 *see also* AAR
"A Leaders Guide to After-
Action Reviews" 30, 134

V

values 71–72

W

Weber, R. 134, 138

Y

Yahoo! 65